The
Arabian
Connection

A Conspiracy
Against
Humanity

Dr. Kasem Khaleel

Knowledge House Publishers
for the Institute of Scientific Wisdon
Lincolnshire, Illinois

Printed in the United States of America

First Edition

This book is a completely revised version of the former title,

Miracle of Islamic Science

ISBN: 0-911119-70-1

For ordering information write to:

Institute of Scientific Wisdom

P. O. Box 855

Lincolnshire, Illinois 60069

To order via voice mail leave name, address, telephone number, and payment information on our secured voice mail: 1-800-295-3737. Books are $14.95 each plus $5.00 S & H. Add an additional $1.00 for each additional book.

Also by the same author : *Incredible Islamic Scientists*, Volumes 1 & 2

The Incredible Qur'an

Acknowledgements

God is Great; There is no other god but He.

Dedication

To all those kind, lovely souls who are/were tyrannized by other humans merely because of who they are. May God release them from the grips of the tyrants and may the decent people enjoy the rest of their years in peace and solitude and in the friendship of others of their kind.

Table of Contents

Foreword

It takes bravery to correct historical fabrications. History has been altered and lies have been perpetrated merely to suit the agenda of those in power. To rock a cultural boat is a dangerous endeavor, since only the enlightened, open minded individual will even care to acknowledge such an effort let alone readily accept and believe it. Anything that isn't status quo is usually vilified and met with fear and hatred. Muhammad was the first and only individual to focus on the accumulated knowledge bank available to humanity and turn unconceived potential into reality. The intense focus on knowledge instilled by Muhammad encouraged Islamic scholars to achieve monumental advancements individually and collectively that have never been matched. His wisdom of human behavior and astute guidance caused a totally barbaric way of existence to become transformed into a civilized, gentile society, which was balanced, tolerant, and thoroughly educated. This apogee of human development fails to exist today. Yet, it could exist simply by adhering to the timeless principles he and other men like him, such as Christ, Moses, Abraham, Confucius, Bhudda, and innumerable others, elaborated.

If this world community is to survive, the current acceptance of lying, cheating, stealing, larceny, and personal gratification at any expense must be halted. To make money at any expense, including fatally harming the earth and its inhabitants, is the epitome of greed. Racism is the poison of society. It justifies the hate and torture—even murder—of others. If such human decadence becomes the norm, there is little hope that humanity can recover. In fact, it will probably be the doom of civilization as we know it.

J. K.

Introduction

Does it really matter who were the originators of modern science? Certainly, to historians it does. Yet, everyone should have at least a basic understanding of who they were. This is because the founders of science are responsible for many of the great advances of the modern world, which humanity benefits from every day.

The majority of individuals believe that the way history is taught is accurate. It is reasonable to presume that modern historians only provide the most accurate data. No one would purposely teach fallacious history. In particular, the authors of history books would conceivably publish only the most correct data, otherwise they would have to reprint their books. Yet, unfortunately, the facts of history are filtered and sifted according to the beliefs and prejudices of the instructors as well as the writers. Indeed, thousands of history books are replete with monumental errors and omissions that disguise the truths of history.

With regard to the history of science a variety of significant errors are being made in educational institutions throughout the world. Inaccurate history is taught and substantive data is omitted. In particular, the history books and encyclopedias on the sciences are overwhelmingly biased against Islamic science. Many of these books leave the impression that the Muslims (Arabs) acted merely as a conduit for the transfer of the wisdom of the ancients to Europe. Few if any original contributions are mentioned. For instance, the *World Book Encyclopedia* gives two sentences to the so-called Arab contribution in the section on the history of science (see *World Book Encyclopedia*, Volume 16, under the heading Science). Additionally, virtually all of the film strips and videos dealing with the history of science completely omit the Islamic contribution.

It is commonly believed that the scholars of ancient Greece originated the principles of modern science. The Greeks ruled the world for 500 years. The Romans were its lords for nearly 600 years. Their empires were massive, the Roman one being the most considerable. Yet, neither achieved the size nor grandeur of the Islamic Empire at its zenith. Nor did they rule as long. Furthermore, their literary and scientific output were incomparably less. Even with the spectacular accomplishments of the Greeks there simply is no comparison to those achieved in the era of Islam. The Islamic scholars thoroughly exceeded Grecian output in the exact sciences as well as in religion and philosophy.[1-10]

During the Roman dynasty the production of literature was minimal. In the Medieval West it was non-existent.[11]

At its pinnacle during the thirteenth century the Cairo library contained some two million books. The library in Tripoli housed an even greater amount. Untold millions of books could be found in the libraries of Spain. Other uncountable manuscripts existed in the public libraries of Baghdad, Mosul, Rayy, Samarkand, Aleppo, Tripoli, and Damascus. Private libraries of rulers, dignitaries, and scholars contained thousands and in some instances hundreds of thousands of books each. The 5 million books which existed within the libraries of Cairo and Tripoli alone represented an incredibly large amount for such an early era. Thus, tens of millions of books were distributed throughout the wide extent of the Islamic Empire. Perhaps even more astounding is the fact that virtually of all these books were hand written. In contrast, today's New York Public Library, the world's largest, houses some 7 million *mass* printed books.

Such a vast literary output exerted a profound impact upon the advancement of civilization. Yet, there is no mention of such an influence in the standard history books and encyclopedias, the references that youngsters and college students throughout the world utilize to prepare reports and study for tests. A sentence here, a sentence there, that is the Islamic or, to put it in the more commonly used verbiage, "Arab" contribution to civilization. Let us again use the World Book Encyclopedia as an example. In its seven paragraph discussion of the history of chemistry not a single word is mentioned regarding the Islamic accomplishments. Concerning the history of geology a total void is found until 1795, when only European achievements are listed. Regarding geography no mention of the brilliant works of the Islamic geographers is made. The editors did mention that the Muslims "knew the earth was round." In geometry several paragraphs focus on the Grecian innovations. A total void is left from 200 B.C. until the 1600s, a period of over 1800 years. Yet, this is precisely the timespan which includes the era of Islamic dominance in this science. In algebra the Islamic contributions are unavoidably given a paragraph, as they are regarded by all scientific historians as its inventors. However, despite this the World Book Encyclopedia mentions Diophantus of the 3rd century A.D. as the "father of algebra." In anesthesia an extensive omission is made, as none of the Islamic developments are mentioned. The astronomy section leaves a near total void between 150 B.C. and 1500 A.D., although the Islamic influence upon Columbus in respect to the roundness of the earth is mentioned. None of the premier astronomers of the Islamic Empire are named. Under the headings *Pharmacology and Pharmacy*, sciences of which the Muslims were exclusively the founders, the omission is blatantly obvious. In medicine, the most extensive field of Islamic accomplishment, not even a hint is given of a major or, for that matter, minor contribution. In nearly two full pages the only comment is that "Arab medicine," among other cultural types, was taught at the University of Salerno.

Modern historians have obliterated leagues of data from the historical record. They have done so motivated by prejudice and special interest. Such an attitude is offensive even to Western historians. It was Paul Benoit and Francoise Micheau who noted a definite bias in Western writings and lectures on the history of science, which they described as an example of "Eurocentrism." They even labeled the seemingly deliberate omissions of the Islamic works as an example of outright racism. These Western scholars confirm that history should be recorded accurately, and the correct credit must be given to whomever is responsible for the discoveries, whether Western, Eastern, or Oriental.

Medieval Europe was largely barbaric and was completely devoid of organized civilization. Yet, the history books offer no mention as to exactly what transformed the archaic Medieval European society from barbarism to modernism. Often, the Crusaders are mentioned, the implication being that their excursions, regardless of how devious, contributed to the advancement of civilization. Yet, the scope of the atrocities perpetrated by the Crusaders is seldom if ever portrayed. Some of these texts suggest that while the Crusades were a dark era in European culture, they served as a channel for the advancement of Western civilization through contact with the Islamic one. Certain authors suggest that the Crusaders were largely responsible for the renovation of European civilization. This is a highly distorted and inaccurate portrayal of that violently murderous, disgusting stage in human history. The Crusaders were motivated only by plunder. The civilization of Islam was instilled with the obligation to enhance humanity, and this is precisely what it achieved.

The majority of individuals believe that the Middle Ages was a period of global decadence. This is based upon the fact that Europe was backward, which is universally acknowledged. Yet, admittedly, few individuals have any clue regarding the status of the rest of the world: Byzantium, Arabia, North Africa, Ethiopia, Persia, China, India, Iraq, and Russia.

The Middle Ages spanned over 1000 years, from 450 to 1500 A.D. This period is also known as the Dark Ages or the Medieval period. The term Dark Ages implies that little of significance arose from human minds in respect to science, culture, or civilization. In fact, it is fallaciously believed that the entire world was immersed in barbarism. The term is accurate only regionally, that is for the continent of Europe west of Spain. Meanwhile, only a few hundred miles from Europe civilization was at its greatest height. During the midst of Europe's darkest ages there existed simultaneously the most massive scientific movement ever known in history.12 Here, under the auspices of Islam thousands of scholars and intellectuals gathered to study the sciences. This movement was so profound, so valuable, so revolutionary, and so intense that Goldstein described it as the West's "Gift of Islam," while Sarton deemed it "The Miracle of Arabic Science." Therefore, the terminology is erroneous and must be redefined, for instance:

Pre-Islamic Era: 400 to 621 A.D.
Dark Ages of Europe: 400 to 1500 A.D.
The Islamic Era: 622 to 1492 A.D.

Those who study history know that numerous magnificent cultures besides the modern ones have existed. These cultures differed in many respects: in beliefs, customs, technology, industry, and language. However, the majority of them, whether ancient or modern, had one element in common; they were intolerant of other peoples. In other words, they exhibited racism.

The rulers of ancient Egypt were racists, tormenting and enslaving untold numbers of peoples of other cultures such as the Jews. The Babylonians of ancient Iraq perceived of themselves as the master race, enslaving and persecuting tens of thousands of people over hundreds of years. Today's Iraqi ruler is more of a thug than a racist.

The ancient Greeks were ethnocentric. They regarded all non-Greeks as culturally and "humanly" inferior to themselves. They coined the word barbarian to describe anyone who couldn't speak Greek. In Greece an individual could only become a citizen through heredity.

The Romans committed unspeakable crimes against humanity in the name of race advancement. They brutalized, in fact, murdered, the followers of Christ, who refused to submit to their pagan beliefs.

The Arabs of Muhammad's era were not accustomed to the racist ideology. Rather, they were known for their unusual and sincere hospitality. This was one of the major reasons they were chosen for bearing his message, because Islam thoroughly condemns racism.

The Colonists of the Americas were racists. They regarded the American Indians as "savages" from the beginning. Spain's conquistadors, notably Cortez and Coronado, were probably the most brutal of all. As described by S. H. Pasha, systematically, the cultures and civilizations of the Indians were destroyed to such a degree that no trace of them remains. That is the ultimate form of racism.13

The unspeakable crimes committed against the American Indians were often done in the name of "civilization" and "Christian" advancement. Yet, this race extermination has in it no trace of Christianity and certainly no foundation in the teachings of Christ, an incredibly gentle man who taught non-violence. It was simply a ruse for the underlying element: white European supremacy. To this day the American Indians are treated as second class citizens. They are continuously subjected to racial slurs.

In the 20th century numerous racist societies have existed or currently exist. Hitler, Stalin, and Mussolini were racists. In modern Britain racism is a policy. Like the ancient Greeks, heredity is a prerequisite for citizenship. The Jews of modern Israel are racists. Their treatment of the Palestinians is inhumane. Bulldozing peoples' homes right from underneath them or stealing them while the inhabitants are away are examples of their blatant discrimination. South Africa, a nation which had perfected modern racism, may have improved in theory, but racism is still endemic there. In North America racism abounds. Some of it is obvious; for instance, obnoxious groups, such as the Neo-Nazis, White Supremists, and the Ku Klux Klan, are still free to "practice." However, usually, racism is more subtle, like the attitude of viewing individuals of different skin colors as a "curiosity." In America individuals with unique skin tones or heavy accents are routinely questioned about their "origins." Furthermore, they are frequently harassed, especially at airports or other federally controlled facilities. Individuals who have a heavy accent and/or dark skin are often relentlessly questioned, "Where are you from?" or, "What is your origin?" Ironically, America, the land of the free founded upon the ideal of freedom of race, culture, and creed, once belonged to a people of dark, beautiful color, although they were derided for it, i.e. "the red skins."

What a superior world it would have become if the red and white skinned people would have worked together to enhance civilization.14 As eloquently described by S. H. Pasha this is precisely what would have logically occurred if the Muslims would have landed in America instead of the Spaniards. The extermination of the Native Americans and with them their valuable culture, agriculture, and medicinal herbs, as well as philosophy, is a profound detriment to Western civilization, one that will haunt it forever.

The modern attitude towards women is perhaps a more subdued example of racial intolerance. For instance, in the job force women of equal if not superior abilities are routinely paid less than men, and this is true of even highly skilled ones such as physicians.15 Yet, perhaps the most obvious example of inequality of women relates to their treatment sexually. In the workplace they are viewed as "sex objects" and are continuously belittled by derogatory comments. They are even physically accosted, a dilemma made evident by the recent (1998) Mitsubishi scandal in Bloomington, Illinois. Here, top officials, as well as factory workers, wantonly abused female workers sexually. It is not only because men are usually stronger "physically," but also because they may hold a position of "power" that they feel comfortable intimidating and/or abusing women at will. In a country which prides itself as the bastion of freedom for women, sex scandals, even in the highest offices, are routine. Women are far from free, unless freedom is defined as free sex, sexual harassment, coercion, and/or rape.

In a truly Islamic civilization racism and sexism would be outlawed. Anyone caught instigating it would be severely punished. Islam has no "tolerance" for racism.

The equality of the races is the cardinal feature of Islam. It is an attribute which advances civilization to its greatest heights. Under its tutelage all races, as well as the sexes, immediately become equal. Thus, exploitation is eradicated. This benevolence towards people of all races and societies, this unbridled equality of all peoples, is the primary difference between the Islamic culture which existed from the 7th to 14th centuries and any other. This was a society where scientific advancement occurred without the desecration of culture, the disintegration of social values, or the destruction of the environment. It was a civilization which promoted and exemplified social tolerance rather than feigning an artificial interest in it, as does the Western world today.

Western countries are poor examples of racial tolerance. In fact, Western civilization breeds racism. Incredibly, it sanctions, in fact, encourages, the existence of racist organizations. In America blacks still commonly suffer racial slurs. Violent acts, such as church and cross burnings, as well as beatings and murders, are instigated simply because skin color. For example, in the recent past (February, 1998) the only remaining black activist newspaper in Mississippi was destroyed by a firebomb. There is a reason racism is encouraged: it is the basis of domination. In other words, it is what keeps Western civilization powerful.

In contrast, in a truly Islamic civilization there would exist equal opportunities for all individuals from all races and cultures, regardless of the career being pursued. Discrimination due to sex, skin color, or race would be unknown. Coercion of any sort would be banned.

As a blatant example of racial intolerance, Western powers constantly foist tyrannical rulers upon the impoverished masses. Then, they keep the tyrants in power through taxpayer money as well as illicit funds. Through this well planned approach, they completely prohibit the people from achieving power and freedom. Thus, they stymie any opportunity for society to progress. The fact is the West has never been a leader of freedom. Instead, through its repressive policies the rights of humans — billions of them, including its own citizens, — have been severely compromised.

In contrast, in the Islamic civilization of the Middle Ages the issue of racial/cultural tolerance never arose. It was standard policy that the races are equal. Said Muhammad, "He who engages in racism is not one of us." By this definition Kuwait, Saudi Arabia, Abu Dhabi, Bahrain, the Arab Emirates, and all other Sheikdoms cannot be regarded as Islamic civilizations, since pride of lineage is a form of racism. Nor can Egypt, Syria, Jordan, Algeria, Pakistan, Iraq, Morocco, or Libya. Rulers there haven't been elected by vote for decades.

Virtually all countries in the world appear to be engaging in racist behavior. The nations of the earth lack a model civilization where men and women of different colors and creeds are treated equally in every respect and where they can work together for the advancement of humanity.

Islam opposes racism in all of its forms. In fact, it seeks to erase it from the globe. During the Islamic Era international freedom and interchange reached such an incredible height that it is yet to be duplicated. Anyone could travel to the hundreds of countries of the Islamic Empire free of harassment and without visa or passport. This is because Islam is interested in only one phenomenon: the advancement of civilization. Islam created civilization in an era wherein there was no civilization anywhere in the world. Furthermore, it continues to create it.16

There is greatness in the United States and other Western countries. However, there is also immense evil. For instance, singlehandedly industrialized countries have polluted the entire globe, causing irreparable damage to wildlife as well as humans. As a result, the future of all subsequent creation is jeopardized.

Islam condemns the wanton desecration of the earth. The fact is the greatness of modern civilization is profoundly enhanced by the grandeur of Islam. The result is an infinitely superior place to live.

The final chapter of this book describes a society which was the model of racial tolerance: Islamic Spain. Possibly, the reader can imagine what it must have been like to have lived in that wonderful world. In this society everyone had equal opportunity to excel to whatever level he/she aspired. It was a civilization in which individuals of various colors, races, and creeds flourished, which had never previously occurred in human history. Incredibly, it was primarily Caucasian society where anyone of any race could have become scholar, scientist, ruler, king, or president.17 In this respect the culture of Islamic Spain was far more advanced than any in existence currently.

References
(see Bibliography for more detailed information)
Note: If a reference is used more than once, it is repeated
by either using only the author's name or the term, *ibid*.

1. Turner, H. R. *Science in Medieval Islam*.

2. Sarton, George. *Introduction to the History of Science*.

3. Bernal, J. *Science in History*.

4. De Vaux, Cara. *The Philosophers of Islam*.

5. Goldstein, Thomas. *Dawn of Modern Civilization*.

6. *Dictionary of Scientific Biography*.

7. Dunlop, D. M. *Arabic Civilization to AD 1500*.

8. Campbell, D. *Arabian Medicine and its Influence on the Middle Ages*.

9. Bammate, H. *Muslim Contribution to Civilization*.

11. Goldstein, Thomas.

10. Nasr, S. H. *Science and Civilization in Islam*.

12. Goldstein, Thomas.

13. Pasha, S. H. Personal Communication.

14. Ibid.

15. Waalen, J. Women in medicine: bringing gender issues to the fore. *JAMA*.

16. Pasha, S. H.

17. Ibid.

CHAPTER 1

Chemistry and the Beginnings of Science

The natural sciences are acclaimed as recent developments in human history. It is commonly believed that they are solely a European invention. The sciences supposedly developed as a direct result of the Renaissance era, when rational thought overcame the dogma of religion. While the Greeks are regarded as the originators of the principles of modern science, the enlightened scholars of the Renaissance and post-Renaissance period are credited with refining these principles into true science. This is what is currently taught in the school systems in the United States and throughout the rest of the Western world.

The advent of works by Boyle and Lavoisier in the seventeenth and eighteenth centuries respectively is thought to mark the beginnings of modern chemistry. The accomplishments of Galileo and Newton in the seventeenth century are regarded as the beginnings of modern physics, while those of Brahe, Kepler, Copernicus, and Galileo in the sixteenth and seventeenth centuries are believed to mark the origin of modern astronomy. However, the facts of history refute the commonly held perception that the beginnings of the basic sciences—chemistry, physiology, medicine, pharmacology, botany, geology, geography, astronomy, and physics—lie with these men, their disciples, or other Europeans. Documentation that modern science is a direct product of the Renaissance and its European scholars is severely lacking. The fact is modern science originated hundreds of years prior to the Renaissance. Although the contributions made by the aforementioned scholars were significant, scientific historians document much earlier developments responsible for the origin of true science as it is defined today.[1-11]

It is astonishing that in the current era, routinely known as the *Information Age*, fallacious history is being taught on a massive scale. The modern world prides itself upon its reliance on truth in science. In other words, the "facts" alone are supposedly the determinant. According to modern thinking anything which can be proven scientifically should be adopted, while, if it is unscientific, if it can't be verified, it should be discarded. Ironically, despite this philosophy the information regarding the origin of modern science within the standard textbooks, the ones read by school children and college students every year, is grossly inaccurate. This is inexcusable, in fact, it is hypocritical.

Regarding the history of science the correct data must be revealed. Historical errors must be rectified. This is particularly important for the primary sciences, the ones that are a standard part of public education. Thus, chemistry, mathematics, medicine, pharmacology, botany, physics, astronomy, geology, and geography will be emphasized. Once the correct information is established, the entire historical record must be refurbished. Let us begin by reviewing a brief summary of the history of ancient knowledge and science.

The Beginnings

Early civilization, that is the existence of organized societies, began some 6,000 years ago in the Middle East. From the banks of the Euphrates and Tigris rivers civilization diverged into various parts of the world. Early man acquired valuable knowledge through practical experiences. Relatively little was recorded in print. Most knowledge was disseminated orally, although a few discoveries and observations were recorded and collected by the inhabitants of Egypt, Babel, Persia, India, China, and Phoenicia. These societies developed a rudimentary knowledge of astronomy, geometry, and mathematics. Most astronomy was, in reality, astrology, although the constellations were mapped, eclipses were predicted, and the movements of the planets were observed. Folk medicine was utilized and certain industrial processes were achieved. However, a systematic approach to science and technology was unknown.

This knowledge was ultimately channeled to Greece, where six to seven centuries prior to the advent of Christ an atmosphere prevailed for the collection and elaboration of data. This marked the beginnings of organized science.

The Greeks were primarily theorizers and contemplators. They pondered over the nature of the world, and this resulted in originality in teachings. The Greeks were productive for some 600 years. Even though their work was primarily theoretical, it was a major impetus in the creation of the sciences. They helped develop a scientific approach to a few sciences, notably astronomy, mechanics, botany, and medicine. Ultimately, Grecian knowledge faltered, especially after the Roman conquest. With Rome in power, a nation which fully neglected scientific inquiry, the advancement of civilization seemed doomed.

Chemistry: One of Humankind's Earliest Sciences

Attempts to understand the chemistry of substances may be regarded as perhaps the most ancient of all the sciences. The earliest record of humanity's interest in chemistry was approximately 3,000 B.C. in the fertile crescent, a time when it was more of an art than a science. Here too is where many of the world's great civilizations, languages, and religions originated.

Tablets record the first known chemists as women, who manufactured perfumes from various natural substances. The ancient Egyptians of the Nile valley were prolific chemists. They produced certain compounds, such as those used in mummification, the chemistry of which remains unknown to this day. By 1,000 B.C. the chemical arts included the making of drugs and dyes, the smelting of metals, and the production of alloys such as bronze. The elements gold, silver, mercury, and lead were refined. The physical properties of certain metals, notably copper, zinc, tin, silver, and gold, were understood. Ancient Egyptians as well as Greeks, Hebrews, Chinese, Persians, and Indians all aided in these developments.

The mystical sciences, now known as astrology and alchemy, developed during this era. The Greeks of Egypt were devout alchemists, constantly attempting to transmute valueless metals, such as iron and lead, into gold and/or silver. In India, where chemicals were divided into male and female, chemists performed yoga during their attempts at transmutations. In the fourth century A.D. Zosimos the Greek described the substance called *Xerion*, a mystical chemical which supposedly changed other metals to gold. Zosimos claimed that by adding a little dab of the magical Xerion to a pile of metal ore in 200 years it would turn into a pile of gold. There was one problem with Zosimos' theory; he forgot to invent a magical chemical of youth so he could live to use the fruits of his labor and prove that it really worked.

This was the extent of the world's insight into chemistry, one that was maintained well into the Middle Ages. While the Greeks had little knowledge of this science, teachers in schools and universities throughout the Western world imply that the origins of chemistry belong exclusively to the Greeks and/or other Westerners. However, scientific historians refute the conception that the scholars of ancient Greece and/or Renaissance Europe developed this science. The fact is when compared to all other

Grecian sciences, chemistry appears to be the most neglected one. The Greeks offered no scientific theories or experiments. Virtually all of their focus was on alchemy. Certainly, minor advancements, accidental or not, may have resulted from their alchemical delving. Yet, the Greeks failed to develop chemistry into a science through the use of scientific methods of study and analysis. In fact, according to Durant the elaboration of "vague theories" was the extent of the Grecian contribution to chemistry.12

Grecian theories regarding the composition and structure of matter were certainly unique. Plato related the universe, the macroworld, to the micro-world of the human body. Aristotle stated that all things in this world could be classified into four groups: fire, which was hot and dry; air, which was hot and moist; water, which was cold and moist; and earth, which was cold and dry. This was the extent of Greek thought in the realm of theoretical chemistry. While they did provide concepts for later thought, these theories were of no value, since they are unintelligible as well as unscientific.

With the rise of the Roman Empire the Grecian developments in chemistry and other sciences were obliterated. The Romans added nothing to the exact sciences. In fact, they completely obstructed the study of philosophy and science, stalling the development of technology and civilization by centuries. When Rome was overrun by barbarian societies (479 A.D.) intellectual stagnation deepened. Humanity had reached a dangerous predicament: the masterpieces of knowledge in the world's greatest libraries were on the brink of being lost or destroyed. In fact, millions of books were burned by the plunderers.

This begins an era known as the Middle or Dark Ages. The barbarian invasion immersed Europe into the abyss of intellectual ignorance. Knowledge was neither propounded nor preserved. The sciences of any kind were unknown. This state of decay worsened by the decade. Yet, during Eurasia's most dire moment an era of brightness beamed from the near-by Middle East. Here, from the 7th to 13th centuries the Muslims were at the height of intellectual achievement. Sarton, the renowned Harvard historian of science, says that concurrent with the decline of knowledge in the West was an era in which the Muslims were intellectually supreme.

The Greeks made valuable contributions for the advancement of science. Their philosophical and theoretical endeavors were thought provoking. Yet, to conclude that modern science owes its existence to them is erroneous. The Grecian contemplators did just that; they sat and pondered and pondered and sat. The result was that they failed to take action on the vast majority of their theories. Furthermore, virtually all of these theories were false. As a result, their books misled scholars throughout the world, impeding the modernization of the sciences and stifling correct scientific thinking for centuries. The Grecian contribution to science is perhaps most succinctly delineated by Pierre Martineau, former Director of Research, *Chicago Tribune*, who said that because of their excessive use of pure logic and contemplation, the ancient Greeks "never developed any real science."[13]

Experimentation is the very basis of the precise sciences. Chemistry, physics, mathematics, and astronomy are all reliant upon it. Just recall the litany of experiments performed by teachers and professors during primary education. Without such an approach, without the attempt to prove or disprove theories, potentially great ideas would remain nebulous, and the sciences could never have become established. Thus, the thrust which led to the development of the sciences could not have come from the thinkers of the "fist on chin" generations. Briffault revealed this in his book, *Making of Humanity*, when he wrote:

"Investigation, accumulation of positive knowledge, minute methods of science, and prolonged observation were alien to Greek temperament. These were *introduced* to Europe by the Arabs. *European science owes its existence to the Arabs* (italics mine)."

This statement is precisely the opposite of what children, teenagers, and adults are taught in the school systems today. Yet, if this is indeed true, as several prominent Western historians insist, then a great disservice is being done to the children and students throughout the world by teaching erroneous history. The school systems and universities are obligated to correct this historical blunder, that being the proclamation that the Greeks and the Renaissance Europeans were the originators of modern science. George Sarton says it's false, Will Durant says it's false, Humboldt says it's false, De Vaux says it's false, and Briffault says it's false. None of these men are Arabs. What's more, these scholars are among the most imminent historians of the Middle Age period. Sarton is regarded as the greatest scientific historian of any time. The implications of their statements are earth shattering. The modern world is teaching a view of history which has been proven faulty by its own top scholars.

How can the history of science be taught as it is today when the most respected scientific historians of the world propound the opposite? How can the editors of World Book Encyclopedia leave, with good conscience, a total void in their "Red Letter Dates In Science" from 650 to 1500 A.D., the era of Islamic science? In fact, this encyclopedia lists only two Red Letter Dates from 100 to 1500 A.D.; Galen's anatomy and physiology in the former date and Leonardo da Vinci's use of the experimental method in the latter. It is well known that the Romans, including Galen, failed to produce significant contributions in the experimental sciences during their reign. However, contributions during the Islamic period were enormous, a fact which is indelibly recorded in the historical record.13

In the Western world fallacious information has been taught regarding the origins of the sciences for over 200 years. It is time that this is changed.

It is realized that the information contained in this book may come as a surprise not only to pupils of history but also for instructors and historians. The projection of the Muslims, as well as Arabic-speaking individuals of other faiths, as the leaders in the advancement of knowledge and science is of immense historical importance. However, the vast majority of individuals are unaware of the connection.

During the Middle Ages the Arab tribes had no interest in the sciences. In fact, they were entirely illiterate. How, then, could such a society become rapidly transformed into the most advanced civilization ever known to exist? It was Sarton, as well as Durant, who described with utter astonishment the phenomenal accomplishments of the Islamic scholars, marveling at their invaluable and lasting achievements in the sciences. Goldstein and Sarton describe this transformation as a "miracle." What is perhaps more miraculous is the fact that only a minority of these scholars were Arabs, the rest being from a diversity of other cultures.

The knowledge provided by the ancients was a great stimulus for Islamic science. Yet, this knowledge was in a state of chaos prior to Islam. Furthermore, the wisdom of the ancients was abandoned, that is no one was utilizing or upgrading it. It was Islam which created a massive revolution in human thought, which impacted humanity globally. This lead to the world's first international scientific movement. No such revolution was generated from ancient Greece. However, to appreciate how this revolution occurred the role played by the Greeks in the development of the sciences must first be reviewed.

The Greeks laid the bedrock for scientific development. However, they left most of their work unfinished, that is it remained in the form of theories. In other words, the majority of Greek authors failed to conduct the scientific experiments necessary to prove their ideas. This is why their books are riddled with cumbersome, confusing theories. What follows are some of their more progressive findings.

In astronomy it appears that they were the first to discover that the earth and other planets revolve around the sun. Much of their knowledge on this matter was probably derived from writings of the ancient Egyptians, Persians, and/or Babylonians. They knew that the earth was round and came incredibly close (within 80 kilometers) of calculating it's circumference. Hipparchus, antiquity's greatest astronomer, charted the heavens and knew the approximate distance between the earth and the moon. He also made a relatively accurate determination of the moon's diameter. Additionally, the Greeks were aware of the orbital motion of the planets. However, most Grecian scholars regarded the earth as the center of the solar system, a belief that was maintained throughout the world for hundreds of years.[14,15]

Around 300 B.C. Euclid utilized the ideas of earlier Greeks and Middle Easterners to state the basic principles of geometry. The Greeks developed the beginnings of trigonometry (e.g. the Pythagorean theorem).

Archimedes, a student of Euclid, continued the master's work and discovered certain laws of physics, including the principle of determining specific gravity.

Hippocrates is regarded as the father of modern medicine. He went beyond established thought to declare that sickness arises from natural causes within the body rather than the wrath of a god. He outlined the Hippocratic oath, which includes the dictum of "above all, do no harm." This oath is taken by medical school graduates to this day. While his philosophy of patient care was prophetic, the majority of his medical theories were erroneous.[16,17]

It is interesting to note that the very existence of Hippocrates has been disputed by some historians.[18,19] The facts of his life are sketchy and are shrouded by legends and fantasy. However, the weight of the evidence appears to support that he did exist and that he worked primarily in the 4th and 3rd centuries B.C. As for his books historians remain divided if not confused as to which were truly written by him and which were merely attributed to his name.

The Greeks made numerous original discoveries in the fields of human anatomy and physiology. Included are the findings that blood is pumped from the heart to the rest of the body, that the pulse is important in diagnosing certain illnesses, and that the brain is divided into several sections, each having its own unique function. They performed a few crude surgeries and used drugs to deaden the pain. Their knowledge of herbal medicine was significant.

These achievements by the Greeks were certainly tremendous considering the intellectual stagnation that existed in the rest of the world at that time and the ancient age in which they lived. However, to announce them as the sole founders of the natural sciences is ludicrous. The Chinese made discoveries of a similar scope and did so in the same era and even earlier. The Babylonians and ancient Egyptians discovered much of what the Greeks knew, and many Greek writers gave credit to the scholars of these ancient societies. Babylonian, Egyptian, and Persian developments in the fields of astronomy, mathematics, and geography significantly influenced Grecian thought on these subjects.

With the exception of geometry and astronomy, Greek discoveries in the precise sciences were relatively few. They disliked intense scientific work. They avoided highly technical or difficult laboratory research. Instead, they thrived upon contemplation and philosophical debate. The fact is compared to other fields of scholarship they produced relatively few precise scientific works. Thus, they failed to systematically use the scientific method on a large scale. In contrast, their writings and accomplishments in the fields of reason, philosophy, and art were voluminous. They were speculators, not scientists. They concentrated on discussing, arguing, thinking, and contemplating. Efforts in the recording of precise observations and detailed data entry were minimal. Remember the words of Briffault: "these were alien to Greek temperament." Then, how did the detail-oriented experimental sciences—physics, analytical astronomy, trigonometry, algebra, arithmetic, and chemistry—arise? What were the origins of the natural sciences, botany, zoology, physiology, meteorology, mineralogy, veterinary medicine, geography, geology, medicine, biology, engineering, and pharmacology? The Greeks contributed to some of these, but only botany and possibly geography could be regarded as their innovations. Obviously, there is a missing link between the works of the ancients and the modernization of science. That link was produced by Islam.

The Rise of Islam

For the purpose of this discussion the terms Middle Ages, Dark Ages, and pre-Islamic Age are synonymous. The time period approximates the decline of the Persian/Roman Empires and the void that remained in civilization thereafter: approximately 450 to 622 A.D. Any growth in the sciences and civilization was stymied during this era. In fact, it was an era of intellectual decay, one which may be referred to as the pre-Islamic Age.

Prior to Islam the world was ruled by illiterate barbarians. As a result, their was no means for civilization to advance. The books, writings, and philosophies of the ancients were neither utilized nor propounded. Education and the pursuit of the sciences was abandoned. The fact is organized, university, and/or public education were completely unknown during that era. Instead, the world was ruled by ignorance, superstition, and chicanery. Law and order were unknown. Human beings had no morals.

In Arabia, that broad peninsula wedged between the Red Sea, Persia, and Palestine, the decay of humanity was exceptionally profound. Countries surrounding it, Persia, Egypt, Syria, and the Byzantine Empire, possessed at least a remnant of civilization. However, Arabia was devoid of even the most minute element of it. This was largely because the Arabian peninsula was isolated from the cultural influences of its neighbors, cut off by vast oceans of sand. Past civilizations saw no interest in piercing the intense desert to disseminate knowledge; they certainly had no interest in conquering it. Arab traders expended all of their energies simply for survival in this inhospitable region. They had neither the time nor energy to devote to the sciences. Those rare individuals who were literate had only a nominal education. Scientists, philosophers, scholars, and/or historians of any type were unknown.

Instead, the Arabs were absorbed with every conceivable corruption. They were crude and licentious. They delved into the occult. They were highly superstitious. Like all barbarians, the Arabs were decadent.

In pre-Islamic Arabia law and order was unknown. The Arabs were maniacal, and minor incidents ignited bloody tribal feuds that lasted for decades. Each tribe claimed the right to rule and, therefore, tyrannize. As a result, theft, arson, rape, and murder were regular events.

Alcoholism was rampant. Sexual promiscuity and prostitution were commonplace. The majority of people smothered themselves in the extremes of these passions.

Perhaps most ferocious was the pre-Islamic Arabs' treatment of women. As in Medieval Europe the women of Arabia had no rights. Hundreds of wives for a single tribal chief was common.

Women were regarded as mere chattel. In fact, they were held on a level less than human. The Arabs considered the status of women so low that infanticide, that is the murdering of new born female infants by burying them alive, was commonplace. The religious experiences of the early Arabs were similar to those of much of the rest of the world. They worshiped stones, trees, stars, and spirits. In short, they revered virtually everything but almighty God. They stooped to the lowest level of human thought by worshipping idols which they themselves constructed: crude statues made of mud and clay, stick-men made of straw and wood. Such was the Arab conception of divine power. Their view of divinity was similar to that of the Greeks, although the latter's brand of idol worship was far more sophisticated.

In the sciences the Arabs produced nothing. In the arts and literature all they could muster were a few verses of poetry, although the Arabic language is sort of a form of art. Thus, the Arabs were totally uncivilized, and this was how they lived for thousands of years.

Islam revolutionized the Arabs within a short 23 years. It raised them from their decadent state to complete mental and social advancement. Idol worship was replaced with the reverence of God almighty. Moral perversion was dislodged by virtue and decency. Within a matter of decades impoverished, illiterate desert dwellers became the world's foremost leaders in spirituality as well as in the sciences and civilization. Plus, they maintained that power for hundreds of years. While modern historians describe the societal impact of various revolutions and cultural revivals, such as the Renaissance and French Revolution, this change in society was the most profound ever recorded in history.[20,21] The Arabs were rapidly transformed from the depths of moral and intellectual stagnation into the models of decency, clemency, sophistication, and scholarship. What's more, this revolution surged far beyond the Arabian peninsula, influencing humanity globally

The revolution was in fact a revolution in thinking. It was a great event for the Arabs and, particularly, for the women of Arabia. For the first time in history Arab men and women contributed to the development of science, culture, and civilization. Yet, to a far greater extent it was a revival for the societies conquered by the Arabic-speaking Muslims. This Islamic revolution halted, in fact, reversed the intellectual decay that permeated the world in that era. The crumbling or lost civilizations of Persia, Syria, Palestine, Sicily, Egypt, North Africa, and Western Europe were all salvaged from intellectual doom.

The Islamic revolution was an enormous benefit for the entire human race. Without it, it is reasonable to presume that there would be no modern world. This is because the Islamic revolution created a striking change in civilization, which permanently altered the course of history. It was the most rapid cultural and global advancement ever experienced in human history.[22-24] It was an event which led to profound developments in the sciences as well as in philosophy and civilization. The stimulation in the growth of knowledge was so monumental that scientific achievement continued unabated for nearly 700 years. What's more, the effects were permanent, since humanity derives uncountable benefits from these achievements today.

Islam is an Arabic word, which means peace. It also means obedience. The peace defines the relationship between each human being and God; it is a peace between all creatures of this world. In other words, Islam is one of the means by which God and human beings work in unison for the betterment of civilization. In Islam, human beings are God's representatives, fulfilling their divinely ordered duties to their creator. In other words, human beings work for God. Ultimately, Islam is that condition wherein the soul of the individual is at peace with God. This is also the means to find peace with fellow humans. Islam is a system of laws.[25] These laws are formulated by God. Their purpose: to teach human beings the difference between right and wrong. In Islam no one is superior to anyone else. Thus, all races and cultures are equal. All of the recipients of the divine law, the prophets Abraham, Moses, Jesus, Muhammad, and countless others, followed the same system of divine regulations. In other words, every prophet since the beginning of time worshiped the one universal God.

Islam revolutionized the pagan Arabs, demanding that they abandon their corrupt practices. It commanded them to cease the worship of idols. It chastised them for tormenting their neighbors. It ordered them to desist from abusing women, from committing adultery, and from failing to give women equal rights. In an achievement unmatched in human history Islam rescued the Arabs from their depraved, miserable condition and elevated them to such heights that in less than 100 years they became the pinnacle of civilization. Prior to Islam Arabian society produced nothing—not even a single reputable individual. As a consequence of Islam the Arabs, and, particularly, the peoples of the societies they conquered produced tens of thousands of scholars, some of which were world leaders in their fields.

Islam's view of the universe and man's role therein is different than the one in vogue today. Here, God has a special relationship with each human being. This is far superior to idol worship, whether ancient or modern. God is constantly active in this world. The concept of a reserved, distant, uninvolved deity is unknown.[25] In other words, God is intimately involved in all of the functions of the universe, the earth, and the human race. Islam's way is that God is directly involved with all of the proceedings of this world, that He is close to man, that all men and women are dear to Him, and that separation from this spiritual-biological connection is impossible.

In Islam all knowledge is directly or indirectly derived from God. Thus, knowledge of all of the sciences arises from Him.

God is the creator of the brain and the human nervous system. He is the originator/designer of the genes and the synthesizer of all biochemical compounds. The spirit, the soul, and the intellect are all His creations. Even the ability to reason comes from Him. God is the producer of everything, whether in the cosmos or on this earth. He alone makes the laws which govern us, for instance, gravity, momentum, evolution, reproduction, regeneration, instinct, and love. Therefore, He is the ultimate source of knowledge in all realms. Those who turn to God maximize their abilities to grasp the truths of the universe and the world we live in.

As a result of these beliefs the Muslims during the Islamic Era viewed God as being the ultimate source of the great truths of this universe. The pinnacle of wisdom, the greatest of all achievements could only be found in His service. The purpose of life is to elevate the self through a serious study of God and His wondrous universe, that is through gaining wisdom and knowledge about the world, knowledge which would benefit the individual as well as humanity. Science is merely a subdivision of this study. This philosophy led the Muslims to regard it as their duty to study and master every branch of knowledge. Thus, they strove to learn everything possible about the universe.

In searching for knowledge Islam strongly recommends that humankind study the sciences, since science itself is regarded as a divinely established system. It is a sensible approach. God creates the elements which are the basis of chemistry; sodium, calcium, potassium, sulfur, magnesium, silicon, aluminum, copper, zinc, etc; He is the inventor of the atoms and the molecular bond, which are the basis of all matter. Thus, turning to God advances and deepens human understanding of the sciences. Thus, the early Muslim scientists had an entirely different view of science than did the ancients or, for that matter, today's scientists. Inspired by this belief the mental giants of Islam produced tens of thousands of books in the sciences, as well as in philosophy, history, theology, and literature, thus changing the course of history forever. It is a miracle that these scientists were finally able to modernize the sciences, which were stifled by the archaic theories of the ancients. Without their efforts Europeans would have never been able to decipher the work of the ancients. Yet, what is most profound is that this was accomplished simply because of a change in belief and because of an intense desire to achieve. In other words, Islamic scientists regarded it as their obligation to produce and advance knowledge.

The basis of their inspiration is the Qur'an: the book of God. The complimentary source is the proclamations of Muhammad, the Prophet of God. Muhammad, may God rest his soul, accomplished two astounding feats. First, he explained the essence of the Book in respect to all fields of human interactions. Second, he provoked the thinking of individuals in various fields of knowledge: literature, economics, industry, philosophy, personal health, and science. Though an unlettered man Muhammad urged his followers to become educated. He demanded that they seek knowledge in all fields. He was the first individual in history to make such a pronouncement. Commandments from the Qur'an added to the urging. This was at a time when virtually all of his followers were illiterate. In fact, no other world leader or revolutionary has ever placed such importance upon education as did Muhammad, may God rest his soul. In contrast, during ancient times rulers discouraged the education of the masses for fear that intellectual development would lead to rebellion. It was Muhammad who taught that one of the greatest deeds a human being could accomplish is to teach an illiterate person how to read and write. During his time prisoners of war were freed after they taught a specified number of people these skills. The fact is his emphasis on the importance of reading and writing, in other words, public education, completely revolutionized human thought. This was at a time when the entire world was immersed in anarchy and barbarism. Thus, the development of the sciences is directly linked to the statements of this one man more so than any other individual in history.

Muhammad promoted the value of learning in all fields for all people regardless of race or sex. Indeed, he was the first to do so on a large scale. As a result, he may be regarded as the founder of public education. Said Muhammad, "Seek knowledge, though you may go as far as China to accumulate it."[26] That represented an incredibly long distance for an individual to travel in that era. This was a trip that would have taken months, perhaps years. The fact that these were the words of a man who could neither read nor write caused them to have an enormous impact upon his followers. As a result, for the first time in history the knowledge of the various ancient civilizations was systematically recovered, translated, studied, and, ultimately, prevented from certain doom.

It must be emphasized that Muhammad was not a chemist, physicist, astronomer, mathematician, or medical expert. He possessed none of the skills or learning of such specialists. He was, above all, a sensible, simple man. Yet, he had an incredible perceptiveness regarding various fields of knowledge, even though he had no practical experience in them. A genius in human development, he realized the importance of motivating individuals to create and accumulate knowledge. Instinctively, he was aware of the profound and permanent impact this would have upon civilization.

Thus, the Qur'an and the preaching of Muhammad caused such a surge in human accomplishment that the world has never known before and has yet to be experienced since. As the Prophet urged the seeking of knowledge in all fields the Muslims of the Middle Ages possessed something unheard of at that time: open minds. They searched for ancient wisdom with a tenacity never before seen in human history. They recovered the earliest sources of the wisdom of human civilization: the writings of the ancient Greeks, Persians, Indians, Egyptians, and Syrians. These works were translated and, where relevant, assimilated. Most of this flurry of translation occurred in a 200 year period spanning from the 9th through the 11th centuries. Incredibly, there was such a drive for the seeking of knowledge that Muslim rulers paid the translators the weight of the book in gold as an incentive.27

These great resources of Hellenistic knowledge were easily accessible to the Muslims, as the storehouses of Greek manuscripts were found in the Near East, not in Europe. For instance, Alexandria, Egypt, and Antioch, Syria, contained a large number of them. So did Constantinople. True, there were some Greek writings to be found in Medieval Europe. However, these lay covered in the dust of monasteries, the keepers of which were too ignorant to understand their meanings or to propound their contents. In contrast, the scholars of the Islamic Empire translated ancient classics and published them on a large scale. Thus, the Muslims performed an immense service for humanity by ensuring that the great works of the ancients were preserved and transmitted to the civilized world.

Chemistry or Alchemy; Which is the Islamic Innovation?

Most of the texts translated by the Muslims dealt with the cult of alchemy. Muslim scholars rapidly discovered that the use of science by the ancients in chemistry was unknown with very few exceptions.

According to the *World Book Encyclopedia* the Grecian word for chemistry, *chemia*, describes "the art of attempting to make....gold and silver from base metals such as lead." Therefore, the term currently used to describe this process, alchemy, which is derived from the Arabic, *al-kimiya*, is a misnomer. Since the Muslim scholars were the first to elevate chemistry from an art to a science, the correct Latin rendition for the modern scientific term should be *alchemiya*, or "alchemistry," while the pagan art should be renamed "chemistry."

According to J.W. Hill alchemy is defined as a mystical chemistry which *"flourished in Europe* during the Middle Ages (italics mine)."[28] Webster's dictionary notes that the chief aim of alchemists was to change non-precious metals into gold. Yet another definition is to discover a secret potion for immortality. By even a loose adherence to these definitions the Muslim scientists of the Middle Ages were not alchemists. Rather, they produced a variety of compounds useful for the development and advancement of science, culture, industry, and civilization. Therefore, they were the first true chemists.[29-35]

The great scientists of the Islamic Era denounced the validity of alchemy, claiming instead that the uniqueness of the elements is created by God. They dealt with the practical applications of chemistry, studying the nature of matter and the properties of various compounds and elements. Then, they applied this knowledge productively for science and industry. They were particularly advanced in using chemistry for the development of pharmacology and medicine. While all civilizations have delved into the occult, it is interesting to note that Islam's top scientists were against this.[36-39] Three of Islam's greatest scholars, al-Biruni, al-Kindi, and Ibn Sina, were diametrically opposed to alchemy. Ironically, the occult sciences, represented by innumerable astrologers, fortune tellers, and satan worshippers, are certainly far more pervasive today than they were in the Islamic Era.

The translation into Arabic of the early works of the Greeks, Egyptians, and other ancient civilizations in the development of chemistry, physics, philosophy, and the remaining sciences was of immense importance. If Europe had her way, these manuscripts would have continued to rot. Many, perhaps the majority, would never have been recovered. Medieval Europe took a dim view of Grecian science. The fact is these works were repudiated in the West as heresy. The Church smothered attempts in the study of the sciences, often burning Grecian books. Indeed, their readers and pupils were themselves frequently burned at the stake. Thus, Europe completely neglected the translation, preservation, and/or elucidation of books on the sciences or, for that matter, philosophy, civilization, history, religion, or any other field.[40,41]

During the Middle Ages Europeans lived in ignorance of the sciences, because the Establishment forbade scientific inquiry. Yet, mysticism and the occult were allowed to thrive. In some regions this intellectual blockade lasted well into the 18th century. As the epitome of intolerance organized Christianity completely repressed free thinking by creating fear in the populous. Furthermore, the wisdom of the ancients was barred from entry into Christian Europe. Thus, all of Europe was held in check by a few bigots. Whatever minimal scientific knowledge which did exist was, once found, systematically exterminated by the ruling class and clergy. Under Constantine public libraries underwent dissolution or were destroyed, while in the 16th century Pope Gregory eliminated scientific studies from Rome, banning the study of all Greek writings.

For hundreds of years learning was branded as magic and treasonous and was punishable by torture or death. Trial was without the right for defense. This was through the so-called Inquisition courts, which reigned supreme from the 12th to 17th centuries. Roger Bacon, the noted English father of science, was considered a heretic and was forced by the rulers of England to abandon his efforts in scientific study; he was imprisoned for some fourteen years, largely because he studied Islamic books. Bruno was burned at the stake for the crime of claiming that the earth was round and that it rotates about its axis.[42]

It was in this, the West's lowest, darkest era of moral and intellectual desolation that the Islamic world brought elegance, intelligence, and brilliance to human civilization. The Muslims preserved the existing scientific and philosophical literature of the ancient world, the works of innumerable scholars over thousands of years, salvaging books, writings, and notations from historical doom. In this regard they were inspired by the dictums of Muhammad, may God rest his soul, that is to seek and preserve knowledge. Furthermore, they produced an immense number of original works, many of which are of greater historical importance than the Grecian texts. Plus, they brought society something far greater than even these: the wisdom of divine revelation and the refreshing, sensible insight of its bearer, Muhammad, may God rest his soul.

After translating the works of the Greeks and Egyptians, the Muslims quickly became aware that in the field of chemistry the ancients dealt primarily with speculation and mystery. The Greeks maintained a constant desire to turn crude relatively valueless minerals into gold, copper, or silver. Metaphysical science clearly dominated fact and experimentation.

The Development of Chemistry

The Muslim scholars who studied the Grecian works found metaphysical chemistry to be unacceptable. They sought to discover the practical uses for chemistry. In fact, historians point to the Muslims as the founders of this science.[43-51]

It must be reiterated that chemistry as a science didn't exist prior to the advent of the Muslims. Certainly, the Greeks, Persians, Chinese, and others understood some of the elements: lead, copper, and iron, for example. However, they knew nothing of the more crucial substances such as the acids, alkalis, alcohols, and tissue salts. Chemistry would fail to exist without these compounds. Furthermore, the ancients were unfamiliar with the use of various processes of chemistry, such as distillation, sublimation, crystallization, oxidation, and precipitation, all of which were invited and/or perfected by the Muslims.52-56 These processes are the very basis of modern chemistry. In addition, Muslim chemists discovered the process of calcination, which is utilized to reduce substances to a powdered form. This technique is invaluable for the synthesis of drugs.57

Muslim chemists were the first to synthesize chemicals on a large scale. They produced a variety of compounds for scientific and medicinal usage. They initiated the synthesis of mineral compounds in order to mimic the processes of Nature: iron sulfate, mercury sulfide, mercury oxide, copper sulfate, copper sulfide, sodium bicarbonate, and potassium sulfide were among their formulations.

It is necessary only to quote Will Durant to comprehend the scope of the Islamic contribution to chemistry. He writes that the Muslim scholars "introduced precise observation, controlled experiment, and careful records. They "invented and named the alembic (*al-anbeeq*)", i.e. the distillation condenser, "chemically analyzed innumerable substances, distinguished alkalis and acids, investigated their affinities, studied and manufactured hundreds of drugs." Further, Durant makes it clear that the mystical alchemy, which is routinely attributed to them, was not their invention. The fact is alchemy, that is the "science" of turning valueless metals into gold or silver, is entirely a Grecian innovation. Yet, hundreds, perhaps thousands, of Western textbooks erroneously attribute it, either directly or by inference, to Islam.

Islamic chemists originated the synthesis of numerous crucial substances, which are essential to the development of the chemical sciences. They include sulfuric acid, nitric acid, mineral oxides, mineral sulfides, hydrochloric acid, sodium hydroxide, silver nitrate, potassium nitrate, potassium carbonate, sulfuric chloride, arsenic bichloride, nitro-glycerine, aqua regia (a mixture of nitric and hydrochloric acids), ammonia, and ammonium chloride.58-60 These reagents are found in virtually all chemistry labs today. The production of these reagents required considerable expertise in the chemical sciences. Judd notes that Muslim chemists operated the "first systematic laboratories" in history.61

The acid-base principle of chemistry is entirely an Islamic development. Therefore, Muslim chemists were the forerunners for the creation of one of the most crucial concepts in this science: the pH scale. Evidence for this is found in the fact that the term alkali, or base, is derived from an Arabic word (i.e. *al-kili*). Sodium hydroxide was the most important alkali synthesized by the Muslims. It is an invaluable reagent for use in modern research and industrial chemistry. Another important base, ammonia, was produced on a large scale from organic material such as hair and blood. In uncovering the mystery of who discovered the mineral acids, which are among the most critical chemicals for modern analytical chemistry, most of the evidence points to the Muslim chemist al-Jabr. Al-Jabr (Jabr bin Hayyan d. 815) became widely known both in Europe and in the Middle East as the father of chemistry. His works are regarded as equal in importance to those of Priestly and Lavoisier.62,63 As early as the eighth century A.D. he presented theoretical concepts which would astound the modern chemistry oriented mind.

Al-Jabr stated that all matter can be traced to a simple, basic particle composed of a lightning-like charge and fire, which serves as the smallest indivisible unit of matter, a concept probably derived from ancient Greek writings. This statement is an early example of the atomic theory, i.e. the electrically charged atom. He is regarded as the discoverer of as many as 19 elements and is credited with the correct measurement of their specific weights. Al-Jabr and colleagues perfected the use of various chemical processes such as distillation, crystallization, and sublimation. He was the first to distill vinegar into concentrated acetic acid, which he used industrially. He also utilized distillation to produce a variety of concentrated reagents such as nitric and sulfuric acids. Al-Jabr introduced the concept of solutions regarding the relative solubility or insolubility of substances. The use of water baths, which is a fundamental part of education in chemistry classes all over the world, is attributed to him, although Islam's famed physician ar-Razi also used them. Like ar-Razi al-Jabr was a physician, and, thus, much of his chemistry was a contribution to efforts in procuring improved drugs. Al-Jabr and other Muslim chemists were the first to use glass bottles (beakers, etc.) and test tubes on a large scale in the study and production of chemical compounds.

Hundreds of books and essays attributed to al-Jabr are found as treasures in the national libraries of Paris and Berlin. Particularly astounding is his *Chemical Compositions*, which remained the authoritative text on chemistry in Europe until the 18th century. The fact that al-Jabr's writings heavily influenced post-Renaissance chemists is indisputable. For instance, Will Durant notes that when al-Jabr's works were translated into Latin during the 13th through 14th centuries, they "strongly stimulated the development of European chemistry." Mathe says they were Lavoisier's "bible."[64]

Al-Kindi, a 9th century Muslim chemist, is probably most deserving of the attribute of elevating chemistry from a cult to a science. He was a strong enemy of alchemy, denying the validity of the Grecian theory for changing non-precious metals into precious ones. He noted it is impossible for chemists to change other metals to silver, saying man can't duplicate the processes of nature. Thus, al-Kindi clearly disputed the "authority of the ancients," since the latter believed in and practiced mystical chemistry. If a concept could not be proven analytically, al-Kindi rejected it. His bravery in this respect must be regarded as epic, a turning point in the transformation of the chemistry laboratory from a witch's brewery to a facility of science. Thus, in the Islamic Era al-kimiya was precisely what it is today: the study of chemicals.65-68

Al-Kindi mastered the use of chemical purification processes such as distillation, sublimation, precipitation, and solutions. He also synthesized a variety of compounds. As a result of these and many other accomplishments al-Kindi may be regarded, along with al-Jabr, as the founder of chemistry as the exact science that we know it to be currently. Personal drawings of his laboratory apparati exist to this day.

Al-Kindi's 9th century findings in chemistry should be regarded as equal to those of Robert Boyle (17th century) and Antoine Lavoisier (18th century). He certainly laid the groundwork for these latter-day geniuses.

The famed physician ar-Razi (d. 925) was yet another Muslim chemist. He too was a true chemist, and he performed experiments primarily as a means of advancing the cause of medicine. His booklet, *Secrets of Secrets*, is said to be the first example known of a chemistry lab manual. The chemicals he experimented with included nitric and sulfuric acid. His laboratory included burners, water baths, a hearth, a kiln, bellows, crucibles, distillation apparati, ladles, tongs, shears, scales, stoves, filters, ceramic dishes, flasks, and vials. These are precisely the devices which are found in today's high school and college chemistry labs. Clearly, ar-Razi's was one of the first true chemistry labs ever to exist. While the "chemists" of Medieval Europe were attempting to produce love potions and immortality elixirs, ar-Razi was at work in his laboratory tediously extracting and refining new drugs.

Just who discovered sulfuric acid, whether it was al-Jabr, ar-Razi, or some other Muslim chemist, remains a matter of dispute. However, there is little doubt that it was an Islamic invention. It was first described in Arabic writings in 1200 A.D., although, clearly, it had been developed centuries prior. Asimov claims that sulfuric acid is the single most important industrial chemical in use today.

Muslim chemists were the first to accurately divide the elements. In this regard one of their advancements was the division of substances into the categories of organic (living) and inorganic (non-living), a system which remains in use today. The organic substances were further divided into those of plant or animal origin. In the field of inorganic chemistry they made the first accurate distinction between metals and their alloys, noting that alloys were only mixtures and not true elements. Furthermore, they were aware that precious metals, such as copper, silver, and gold, were distinct elements, although they determined that in nature they are tainted with other elements. This is evidenced by the fact that the Muslims were the first "electrochemists." In this respect they produced pure copper by running a copper sulfate solution over iron filings.[69] Additionally, they made the original discovery that non-precious metals become oxidized when exposed to air, no doubt a phenomenal observation in so early an era.

Muslim chemists made a major contribution to the element theory, which states that everything in this earth consists of 102 basic elements. They discovered and isolated some 20 elements, which amounts to approximately one-fifth of the periodic table. One of their discoveries, the mineral potassium, still carries its Islamic identity. Its symbol, "K," is derived from the first letter of the Arabic word, *al-kali*.[70] What's more, the word sodium, as well as soda, i.e. *suwaid*, is Arabic derived.

As industrial chemists they applied advanced techniques for extracting minerals and metals. They perfected glass making and introduced to the West the technology for blowing glass and coloring it with metal oxides. They wholly invented glass crystal and developed methods for cutting it with precision. Furthermore, stained glass, as well as window panes, is their innovation.[71,72] Steel production was introduced and perfected by the Muslims. For instance, the steel swords produced by the artisans of Damascus and Islamic Spain were prized throughout the world, because of their quality as well as design. However, their most impressive accomplishment in industrial chemistry was the production of paper.

The introduction of paper is often attributed to the Chinese. It is true that the Chinese produced paper, but this was primarily through a tedious process using silk. It was the Muslims who first manufactured paper on an industrial scale.[73,74] Thus, they established the world's first paper mills. Later, they made this industry available to the West. The tens of millions of books they produced during the Middle Ages clearly confirms this.

The systematic production of large quantities of paper occurred as early as the 9th century. Wood was scarce in the Middle East. Muslim ingenuity solved that dilemma, since paper was produced from cotton, linen, flax fibers, hemp fibers, and rags. However, in Spain, which did have wood, an entirely different resource was used as a raw material: esparto grass, which was grown near the sea-side region of Valencia in eastern Spain.[75]

The importance of paper production is difficult to quantify. It opened a new era for civilization. Slowly but steadily, Europeans became accustomed to the luxury of imported Islamic paper. In the 12th century Roger II of Sicily signed his first paper document.

The diffusion of inexpensive books and paper products in Medieval Europe only became possible because of the replacement of the cumbersome parchment and silk paper. Eventually, the Europeans, following the Islamic model, constructed their own paper mills. Certain historians regard the introduction of paper milling to the West as the single most important technological development of the Middle Ages. The availability of cheap paper, which was later combined with printing machines, revolutionized the Western world, which was suffocating in the coffin of illiteracy.

The Muslims initiated the large scale use of a number of substances utilized in construction/architecture. They had a thorough knowledge of industrial chemistry, since they produced a variety of plasters, glazes, mortars, brick, varnishes, and other building compounds which were the legacy of Islamic civilization. Islamic Spain boasted roads paved of cement instead of the cumbersome substances used previously, i.e. concrete and stones, as in the Roman highways. The roads in the cities of Spain were wide, almost like highways, and were illuminated with the world's first street lamps. While the chemical process by which they produced light remains unknown, it is likely that they were fueled by kerosene, since Muslim chemists distilled it from crude oil on a rather large scale.[76,77]

Muslim chemists perfected the technology for producing and/or extracting various natural coloring agents (dyes) for use in tiles, woodworking, and clothing, a technology which remains currently unmatched. It is important to note that the production of clothing from cotton (from the Arabic, *qutn*) was their invention, and they developed a variety of ingenious methods for refining it, improving its texture, weaving it, and coloring it. In fact, the ever popular cotton clothing we wear today, from pants to underwear, has its origin in Islam.

The invention of gunpowder has long been linked to names like Roger Bacon, Albertus Magnus, and Berthold Swartz. It is also attributed to the Chinese. However, research by Reinuad and Favé has clearly shown that, "though the Chinese discovered saltpetre, it was the Arabs and Arabs alone who discovered gunpowder as an explosive substance capable of firing projectiles," i.e. who invented firearms.[78] Furthermore, they were the first to use firearms and other explosives in warfare, notably to defend Algeciras when it was attacked during the 14th century by Alphonse the 11th (Algeciras is a port in southern Spain on the Straight of Gibraltar). The fact is the first wheeled cannons were utilized by Muslim armies during the 16th century. They were one of the secrets of Sulaiman the Magnificent's decisive victories.

It must be emphasized that Muslim chemists synthesized nitroglycerine as early as the 10th century. Gunpowder, as an explosive compound for military use, was unknown in the Western world prior to Islamic rule. While it was the Chinese who invented saltpeter, which is essentially potassium nitrate, primarily for fireworks, they realized no military value. Apparently, Muslim travelers took note of this. However, Durant reports that the Muslims were mining saltpeter as early as 870 A.D. Later, Muslim chemists added sulfur, carbon, and other chemicals to produce gunpowder. They realized its tactical/military usage and produced the first firearms and grenades. Roger Bacon (England, 13th century) imitated the formula after reading translations of Arabic texts and introduced it to the West.

Because of the monumental efforts of Islamic scientists in the field of chemistry, modern humanity continues to receive uncountable benefits. Their legacy includes industrial paper production, gunpowder/firearms, colored and art glass, stained glass windows, fine mortars and plasters, the domed arch, vaulted ceilings, fine perfumes, distilled essences, drug synthesis, and distilled water merely to name a few.

The Muslims' systematic usage of distillation completely revolutionized the science of chemistry. They invented distilled water, which they used in medicine and industry. They were also the first to produce pure alcohol, which was made by distilling fermented carbohydrates. The essential oils of plants were produced on a large scale: al-Kindi described at least 60 different ones.[79] They also distilled aromatic waters, using them as perfumes and medicines.[80-82] A partial list of the substances produced by Muslim chemists through distillation and other chemical refinement processes includes:

• drugs (primarily from medicinal herbs)
• tinctures (from medicinal herbs)
• aldehydes (e.g. formaldehyde, acetaldehyde)
• rose oil   • sulfuric acid

- jasmine oil
- violet water
- violet oil
- ammonia
- orange oil
- lemon peel oil
- rosemary oil
- synthetic musks
- rose water

- hydrochloric acid
- nitric acid
- acetic acid
- orange blossom water
- ammonium chloride and hydroxide
- nitroglycerine
- kerosene and naphtha
- potassium carbonate
- sodium hydroxide

As is indicated by the aforementioned list during the Islamic Era natural plant extraction was extensive, far more so than in any previous civilization. While the Romans used tedious methods that allowed only tiny amounts to be produced yearly, the industrialists of Islam distilled aromatic oils and waters on a monumental scale. This fully illustrates the fact that the now common use of flower oils and essences originated from Islam. Interestingly, the popularity of such substances was even greater in the Islamic Empire than it is today. Rose oil distillation illustrates just how massive their production was. Despite the fact that it takes thousands of pounds of rose petals to produce a few ounces of oil some 30 thousand bottles of rose oil were produced by the Caliphate in one year alone.

The Scientific Method: Western or Islamic Invention?

When students first begin studying the sciences, emphasis is repeatedly placed upon the role of the *experimental method*. Instructors emphasize the critical importance of this method in the creation of the sciences. They stress how its invention altered the course of history. It did so, they say, by forcing scientists to be critical, to rely on data and hard facts rather than presumptions and theories. Thus, the experimental method caused scholars and scientists to test their theories in order to prove or disprove them. As compared to other disciplines chemistry instructors often place the greatest emphasis upon the importance of the scientific method. This is because the science of chemistry is essentially continuous experimentation. Yet, the instructors are unable to provide students with definitive data as to who originated the scientific method. They know it wasn't the Greeks. Usually, they attribute it to post-Renaissance scholars.

It was the Muslims and the Muslims alone who invented the scientific method, perfected it, and introduced it to the West.[83-87] The efforts of Europe's scientists, what few existed, were constrained by illiteracy, intolerance, and religious dogma. In contrast, freedom of scientific expression abounded in the Islamic Empire. The French historian Sedillot states that the Muslim scholars had already thoroughly understood and practiced the scientific method by the 9th century. In fact, for the first time in history hundreds of scientists from all over the world gathered in the universities and observatories of the Empire specifically for furthering the sciences. As a result, an era of scientific achievement began that was so unbelievably productive that George Sarton and Max Vintejoux both described it as a "miracle." Hundreds of thousands of books on the sciences were generated in what is acclaimed as the most voluminous output of original scientific data ever known to occur.

Understanding that true science rests on practical study and experimentation, Islamic scholars studied the works of the ancients with a critical eye. They tested the findings of the ancients and rejected data that failed to adhere to physical laws. Mistakes were corrected and new data was added. Plus, the Muslim scientists performed original research based upon these findings and also upon their own independent theories. Furthermore, they regarded all theories as untenable unless they were proven by scientific studies. This is the scientific method.

It was in Baghdad's prestigious *Bayt al-Hikma* (House of Wisdom) that these developments were initiated. There, in the early 800s Muslim scholars from all over the world gathered to discuss the sciences. They developed the scientific method and perfected it by performing precise experiments based upon the theories of the ancients and those theories formulated as a result of their own independent thinking. In contrast, hundreds of history textbooks credit Roger Bacon of England (13th century) with fathering the scientific method. Bacon learned of experimentation and deductive reasoning through Latin translations of Islamic works. He conducted no significant scientific experiments.[88,89]

Most books on history entirely omit the Islamic developments in creating the scientific method. A typical example is W.S. Fowler's, *The Development of the Scientific Method* (Pergamon Press, 1962.) In his book Fowler implies that the Greeks were the forefathers of the scientific method *even though he admits that they were primarily observers and did not conduct precise experiments.* As for the actual founders of the method he lists Roger Bacon, Francis Bacon, Bruno, Tycho Brahe, Copernicus, Isaac Newton, Leonardo da Vinci, and Galileo. None of the Islamic experimenters are mentioned. Nor did Fowler discuss the fact that each of these men was heavily influenced by the writings of Islamic scientists through their Latin renditions. In addition, he indicates that neither Bruno nor Roger Bacon conducted experiments. Leonardo da Vinci was a major proponent of the scientific method, but he too failed to conduct experiments. How could he have time for it? He was engrossed in his paintings.

According to George Sarton Islamic science had its own "Leonardo da Vinci" five hundred years prior. It was the 11th century Islamic physicist/astronomer al-Biruni. However, al-Biruni was no mere artist or observer. This Persian genius of 1000 A.D. conducted precise experiments on the laws of gravitation, momentum, and motion long before Galileo or Newton. Besides being the author of over two hundred books and treatise, the edition of thousands of books are to his credit. In contrast, da Vinci left the world his paintings, drawings, and "erratic" notes.90 Furthermore, Boorstin notes that while Leonardo intended to write books, "he never published any of them."91

Al-Biruni perfected the scientific method. Among his successful experiments were the most precise measurement of the specific gravity of precious stones and metals, mathematical calculation of the earth's radius, mathematical determination of the earth's solar orbit, and the measurement of the height of mountains via seconds and degrees. Despite these monumental accomplishments he is omitted from Fowler's list. Fowler should have devoted an entire chapter to al-Biruni, as his literary accomplishments exceed those of da Vinci, Galileo, Copernicus, and Newton combined.

The systematic, organized study methods used by the Islamic scholars were based upon a philosophy previously unknown to the world. They believed that the universe had an intelligent order, that it operated in a systematic fashion, and that man could harness this for his own use. Ibn Sina was convinced of it, as was ar-Razi, al-Biruni, Ibn Rushd, al-Haytham, and al-Kindi. This "scientific method" of thinking was understood to modern perfection by the Islamic scientists.

A statement by al-Farabi, one of Islam's most respected philosophers/scientists, further defines the immense impact of Islamic science upon the modernization of the world. He said, "It is our duty to keep silent when we do not know." This philosophy is a profoundly valuable addition to the scientific method. Is al-Farabi proclaiming that, for men of research, to remain silent is an obligation rather than to disseminate questionable science and unproven theories? If so, he was yet another genius of the Islamic Era who was centuries ahead of his time.

Because Islamic scholars were hesitant to publish erroneous ideas, their works were in general highly scientific and were devoid of the cumbersome theories and dogma found in the Grecian writings. Because their writings were practical, they were easily understood and assimilated by Western scholars, who found the Greek writings stultifying. If only the Greeks would have adhered to such a philosophy, civilization would surely have achieved an even more advanced state than exists today.

References

(see Bibliography for more detailed information)

1. Arnold, T. and A. Guillaume. *The Legacy of Islam.*

2. Bammate, H. *Muslim Contribution to Civilization.*

3. Durant, W. *Age of Faith.*

4. Garrison, F. H. *An Introduction to the History of Medicine.*

5. Hitti, P. K. *History of the Arabs.*

6. Turner, H. R. *Science in Medieval Islam.*

7. Sarton, George. *Introduction to the History of Science.*

8. Ronan, C. A. *Science: Its History and Development Among the World's Cultures.*

9. Sedillot, L. A. *History of the Arabs.*

10. *Dictionary of Scientific Biography.*

11. Goldstein, Thomas. *Dawn of Modern Civilization.*

12. Durant, W. *Age of Faith.*

13. See Durant, Sarton, Ronan, *Dictionary of Scientific Biography*, Goldstein, and Turner.

14. Singer, Charles. *A Short History of Scientific Ideas.*

15. *World Book Encyclopedia*, Vols. 1 & 7.

16. Lyons, A. S. and R. J. Petrucelli. *Medicine: An Illustrated History.*

17. Welty, P. T. *Human Expressions: A History of the World.*

18. Lyons, A. S.

19. Welty, P. T.

20. Sarton, George.

21. Goldstein, T.

22. Maududi, S. A. *Towards Understanding Islam.*

23. Sarton, George.

24. Hart, M. H. *The 100: A ranking of the Most Influential Persons in History.* New York: Hart Publ. Co.

25. Pasha, S. H. Personal communication.

26. *The Definitive Statements of Prophet Muhammad.*(Sahih al-Bukhari).

27. Pasha, S. H.

28. Hill, J. W. *Chemistry for Changing Times.*

29. Durant, W.

30. Sarton, G.

31. Singer, C.

32. Ronan, C. A.

33. Butterfield, H. *The Origins of Modern Science.*

34. *Encyclopedia Britannica.* 1973. Vol. 1.

35. Turner, H. R. *Science in Medieval Islam.*

36. *World Book Encyclopedia*, Vol. 7.

37. Dunlop, D. M. *Arabic Civilization to AD1500.*

38. *Dictionary of Scientific Biography.*

39. Durant, W.

40. Braudel, J. D. *The Mediterranean and the Mediterranean World in the Age of Phillip II.*

41. Durant, W.

42. Pasha, S. H. and Singer, C.

43. Bammate, H.

44. Durant, W.

45. Faber, E.

46. Garrison, F. H.

47. Nasr, S. H.

48. Sarton, George.

49. Singer, C.

50. Dunlop, D. M.

51. Dampier, S. W.

52. Turner, H. R.

53. Durant, W.

54. Nasr, S. H.

55. Sarton, G.

56. Singer, C.

57. Scott, S. P. *History of the Moorish Empire.*

58. Bernal, J. D. *Science in History.*

59. Singer, Charles. *A Short History of Science.*

60. Sarton, George.

61. *Dictionary of Scientific Biography.*

62. Judd, G. *A History of Civilization.*

63. Sarton, George.

64. Mathe, Jean. *The Civilization of Islam.*

65. Bernal, J. D.

66. Hill, D. R. *Islamic Science and Engineering.*

67. Durant, W.

68. Singer, Charles.

69. Mansfield, Peter. *The Arab World.*

70. Singer, Charles.

71. Ahsan, M.M. *Social Life Under the Abbasids*

72. Hill, D. R.

73. Durant, W.

74. Sarton, George.

75. Irving, Thomas. *The End of Islamic Spain.*

76. *Encyclopedia Britannica* (under the heading *Petroleum*).

77. Hill, D. R. *Islamic Science and Engineering.*

78. Bammate, H.

79. *Dictionary of Scientific Biography.*

80. Durant, H. and Sarton, G.

81. Ahsan, M.M.

82. Garrison, F. H.

83. *Dictionary of Scientific Biography.*

84. Hill, J. D.

85. Sarton, George.

86. Garrison, F. H.

87. Durant, W.

88. Ibid.

89. Sarton, G.

90. Boorstin, Daniel. *The Discovers.*

91. Ibid.

CHAPTER 2

Physics and Astronomy

It is necessary only to quote Humboldt to demonstrate the origins of physics. He states, "It is the Arabs who should be regarded as the real founders of physics." He makes this claim even though the vast majority of Islamic treatise on physics have been lost. Yet, the small number of works which have survived give full credence to Humboldt's assertion.

It was al-Biruni during the 11th century who paved the way for Newton's laws of gravity. In fact, evidence exists that he invented the First Law 500 years before Newton. Al-Biruni disagreed with the Greeks in the matter of the earth's gravitational pull. The Greeks were correct in noting that heavy objects are drawn to the center of the earth. However, they incorrectly conceived of ethereal substances, such as fire and air, as gravitating toward the heavens. Al-Biruni corrected them, stating "all elements converge on the center of the earth, but the heavier precedes the lighter.."1 Thus, it appears that al-Biruni, not Galileo or Newton, was the first to proclaim that all particles, no matter how large or small, heavy or light, are attracted to the center of the earth through gravitational pull. Furthermore, he believed that the moon is attracted to the earth through this mechanism, as is the earth attracted to the sun. Other Muslim physicists, notably Ibn Bajjah, updated al-Biruni's findings, showing that when the resistance of air is removed, all objects fall towards the earth's center at the same speed as the result of gravitational pull. This is one of the primary laws of gravity.

Al-Biruni achieved his immense scientific contributions primarily during the 11th century. He is regarded by European historians as one of the top 20 scientific geniuses of the Middle Ages. George Sarton calls him "The Leonardo da Vinci of Islam." However, this distinction is insufficient. Al-Biruni preceded da Vinci by more than five centuries and accomplished far more in terms of original scientific input. He wrote and edited hundreds of volumes, made original discoveries, and conducted hundreds of highly precise scientific experiments. Da Vinci contributed only unintelligible notes. He conducted no scientific experiments. Even a reversal of Sarton's words, that is that da Vinci is "The al-Biruni of Christianity" is suspect, because, as Sarton himself makes evident, al-Biruni was a scientist in the purest sense, whereas Leonardo was primarily an artist. Da Vinci's unpublished notes, fine paintings, sculptures, sketches, and anatomical drawings, are insignificant in the realm of scientific contribution compared to al-Biruni's performance of hundreds of scientific experiments and the writing of tens of thousands of published pages on the sciences. Furthermore, while da Vinci's expertise was primarily in the fields of engineering and architecture, al-Biruni was multi-talented *in the field of the exact sciences.* He was an accomplished physicist, chemist, geologist, astronomer, mathematician, biologist, pharmacist, physician, historian, and author. Da Vinci left the world with his paintings. Al-Biruni changed the world through scientific advancements.

As a physicist al-Biruni gave the most precise description for the measurement of specific gravity, one that equals the current method. His method is as follows: "First, weigh the body in air, then weigh it in water. Then weigh the water displaced by the body. From this weight the new weight of the body is found. Then, by dividing the weight of the body in air by the weight displaced in water, we find the specific gravity of a body." This is precisely the method in use today. Al-Biruni also made the discovery of the finite nature of matter. While he knew its forms could be changed, he propounded that its total mass remained the same. This spectacular finding is erroneously attributed to Lavoisier (18th century).

The science of optics, which is a branch of physics, is entirely an Islamic creation. It was al-Haytham's 11th century *Book of Optics*, which M. Charles contends marks "the beginning of the modern science of optics."2 Not content with Euclid's impractical theories, al-Haytham was the first to thoroughly and accurately describe the function of the eyes and its lenses. He also was the first to correctly describe mathematically how the eyes work together to create a visual image, that is he discovered binocular vision.

Al-Haytham must be regarded as the inventor of artificial lenses as we know them today, although Ibn Firnas of Islamic Spain constructed eyeglasses prior. Al-Haytham manufactured both convex and concave lenses, creating a special machine to grind them. The lenses were used to increase the magnifying powers of light. According to Singer and others the magnifying glass is exclusively his invention. His scientific, detailed description of the various lenses he constructed and the magnifying glass was the stimulus for the work of Western scientists in developing the microscope and telescope.

Al-Haytham was the first individual to master the study of light. He was also the first to provide a careful and relatively accurate description of the function of the eyes. This was at a time when the West had no clue as to the mechanism of vision. His works opened the eyes and minds of scholars throughout Europe, stimulating their interest in a wide range of scientific, industrial, architectural, and artistic endeavors. The concept of perspective, which is instrumental in art and architecture, was innovated by him. The fact is virtually all of Europe's famous Renaissance and post-Renaissance artists/architects, such as Toscanelli and Leonardo da Vinci, studied and relied upon the works of the great al-Haytham.

It was al-Haytham, not Newton, as is commonly believed in

the West, who was the first detail oriented scientist. In optics and physics he innovated the use of complicated scientific experiments. His experiments were carefully constructed, plus they were detailed and precise. Furthermore, he applied the scientific method throughout. Therefore, al-Haytham must be regarded not only as the founder of modern optics but also of the entire field of modern physics.[3,4] It was A. O. Williams of Brown University, author of *History of Physics*, who noted that despite Euclid this science was not a Grecian discovery, since "organized...experimentation (in ancient Greece) was unknown." Furthermore, while European pioneers, such as Galileo and Newton, performed scientific experiments, their work was considerably less detailed than the enormously exacting experiments of al-Haytham.[5] What's more, the publications of the European pioneers were largely inspired by the scientific works of Islam, which were translated en masse into Latin during the 12th-14th centuries. The fact is both Galileo and Newton, as well as Kepler, quoted the great al-Haytham. However, Williams, writing in *Americana Encyclopedia*, is apparently biased, because, while completely omitting the Islamic contribution, he notes in rather vague terms that Galileo is "usually regarded" as the father of modern physics. He upholds Galileo on the following bases:

a) that Galileo, who worked in the 17th century, was the first to use observation proven by mathematical analysis
b) that Galileo first correctly described the laws of falling bodies
c) that Galileo invented the pendulum
d) that Galileo founded the basic laws of gravity

Yet, it was al-Haytham who originated the use of detailed mathematical analysis to prove observation long before Galileo (11th century vs. 16th century). Al-Biruni first detailed the laws of falling bodies and the basic laws of gravity 500 years before

Galileo. Ibn Yunus (12th century) categorically invented the pendulum. Therefore, by William's own criteria the Muslims, not Galileo, were the real founders of physics.

Al-Haytham's books were studied in every major Western university throughout the Middle Ages. What's more, they were in vogue for over five centuries. Thousands of European scholars were influenced by them. They marvelled at his descriptions of artificial lenses and the experiments he conducted with them. His discussions of visual projection, that is how the viewed image is mathematically projected from the human eye to the outer world, were crucial to the advancement of science. Without this principle the invention of revolutionary devices, such as microscopes, magnifying glasses, and telescopes, could not have occurred. Thus, al-Haytham's physics singlehandedly changed the world's understanding of light and vision, bringing the brilliance of a high level of sophistication to a world enveloped by darkness and gloom.[1]

This Middle Age genius performed in-depth studies on the physics of how light interacts with the air in our atmosphere. Al-Haytham was the first to discover the reason that the sun and moon appear larger when they are near the horizon. He discovered that this was due to the angular reflection of light rays within the atmosphere.6 He also discovered that light from the sun reaches the earth even when it remains below the horizon as much as nineteen degrees. This finding lead him to calculate the height of the earth's atmosphere at ten miles, i.e. 52,000 feet. His measurement proved to be astonishingly accurate. According to World Book Encyclopedia the outer limit of the first layer of the

---

[1] Al-Haytham's Latin name is still used in modern physics with the so-called "Alhazen's problem." This is described as follows: "In a convex mirror, spherical, conical, or cylindrical, to find the point at which a ray coming from one given position will be reflected to another given position." Al-Haytham solved this using an equation of the fourth degree by means of a hyperbola.

atmosphere is approximately ten miles. Additionally, he made the incredible space age discovery that objects become lighter as the density of the atmosphere thins.

Al-Haytham proposed new theories on how light originates from its source and how it is received and processed by the eye. He was the first to establish correct theories for refraction, that is the bending of light, as well as its subsequent formation into colors, although this discovery is routinely and erroneously attributed to Newton. His experiments with refraction were conducted through every conceivable medium, including air, water, and glass. Through his experiments on the effects of light on mirrors he discovered the Second Law of Reflection.[7,8] Furthermore, al-Haytham invented the correct concept of "rays of light" hundreds of years before it was conceived in the West.

This eleventh century scholar synthesized the most accurate theory of vision ever proposed in the ancient world. He knew that a key to sight was that a visual image is formed in the brain, where the precise size, color, and depth of an object was determined. He correctly realized that there must be a method by which the visual apparatus, that is the eyes, optic nerve, and visual centers within the brain, could perceive the exact distance, shape, and size of the object in view so that the image could be "sent back" to the environment. This lead him to construct a mathematically correct model for visual projection.

Al-Haytham was a prolific writer, producing more than sixty books. Unfortunately, only a few have survived. George Sarton, the renowned Harvard historian, calls him "the greatest Muslim physicist and one of the greatest students of optics of all times." Will Durant says his treatise on optics was "the most thoroughly scientific" of all Middle Age manuscripts. However, it was A. I. Sabra, in the *Dictionary of Scientific Biography*, who best described his immense contribution, proclaiming that what al-Haytham discovered about vision "had not been grasped...by any writer on optics, ancient or modern."

Al-Haytham's writings were a major stimulus for Johannes Kepler's work in astronomy and physics, which World Book Encyclopedia claims "established astronomy as an exact science." However, no mention of al-Haytham's immense contribution is given.

Muslim physicists made enormous inroads in yet another key field of modern physics: projectile motion. It was Ibn Bajjah, famously known in Medieval Europe as *Avempace*, who made the contribution. Ibn Bajjah debunked the ancient Grecian formula for projectile motion, providing instead one which withstood scientific scrutiny. The Greeks believed that a projectile, like an arrow or cannonball, maintained a steady speed when acted upon by a persistent force, whereas Ibn Bajjah proved that this was incorrect and that, instead, the speed is accelerated by a constant force. He, not Galileo, as is commonly believed, developed the now famous formula: $V = P - M$, where V is velocity, P is motive force, and M is resistance. Yet, currently, the full credit for Ibn Bajjah's discovery is awarded to Galileo, who, according to H. C. Corben, plagiarized the formula in his book, *Pisan Dialogue*.[8]

During the Islamic Era the Muslims became highly advanced in the discipline of mechanics. They developed certain principles of mechanics beyond those delineated by the Greeks. They also invented a number of highly specialized mechanical devices, which were primarily agricultural and industrial. Al-Khazini (12th century) created a balance scale which was extremely precise. His scale measured the weight of objects within micrograms, a level of precision surpassed only since the 20th century. The tremendous advances made by al-Khazini as a physicist caused the editors of the monumental work, *Dictionary of Scientific Bibliography,* to proclaim him "among the greatest of any time." Furthermore, they note that the branch of science developed by al-Khazini and his associates, that is precision hydrostatics, is, regrettably, now extinct.

Muslim engineers invented clocks and introduced the technology to Europe. One of the earliest introductions of these devices to Europe was when a Muslim dignitary presented a mechanical water clock to Charlemagne in the 9th century. E. Bernard of Oxford University notes that the Muslims discovered the pendulum for use in clocks, which increased their precision greatly.9 Unfortunately, students all over the world are erroneously taught that this discovery belongs to Galileo. Furthermore, the precise measurement of time, a critical contribution to physics, is an Islamic innovation. Muslim scientists were the first to produce reliable clocks and watches. According to Durant Islamic Spain's Ibn Firnas invented a chronometer, which Britannica defines as a "highly accurate device for measuring time", the first such device ever produced in history.

A most famous Islamic clock was found in the Mosque of Damascus during the 12th century; European observers marveled at its spectacle. Some clocks were weight driven, others driven by windmills, and still others driven by water. Many of these clocks were elaborate, complicated devices and were truly works of art. In one such clock mechanical birds discharged pellets from their beaks at the hour, which fell on cymbals to sound the time change.

Other Islamic accomplishments in physics include the first correct account for the cause of the refraction of white light into the colors of the light spectrum. This was achieved when Kamal ad-Din (13th century), a student of al-Haytham, gave the first satisfactory account of the cause of rainbows as well as the origin of the double rainbow. When there are two rainbows the order of the colors in the second one, a phenomenon previously unexplained, is always reversed. To ascertain this Kamal ad-Din used as his model a series of experiments of light transmitted through glass spheres filled with water. His conclusion was that in the formation of rainbows the light from the sun enters the water drops (from rain) and is then reflected at the far sides of these drops to the observer. The primary rainbow is produced by one such reflection. The secondary rainbow is produced by two internal reflections. Thus, the reversal that occurs in the order of the colors is explained.[10] This laid the groundwork for Isaac Newton's famous discovery in the late 17th century concerning the cause of the colors' production.

There appears to be little doubt that Newton was influenced by ad-Din's and al-Haytham's experiments. Kamal ad-Din's work on rainbows was copied verbatim by Europeans in the 14th century, and to this day his discovery is wrongly attributed to them.[11,12]

Additionally, Kamal ad-Din capitalized on work done by al-Haytham regarding the operation of the camera. The latter had already conducted experiments in a darkroom in which he studied daily positions of the sun as well as eclipses. The unanswered problem was how to explain the formation of inverted (upside-down) images of shining objects, such as the eclipsed sun or moon, inside the dark chamber. The ancients also may have observed this inverted image but left no record of it. While never entirely solving this problem, these Islamic physicists clearly laid the groundwork which allowed for the discovery of the camera many centuries later.

The team of Kamal ad-Din and al-Haytham made scientific history by being the first to observe and record the camera obscura phenomenon. This is literally defined by Webster's dictionary as "dark chamber." This is the pinhole camera, usually consisting of a piece of cardboard with a tiny hole, that grade schoolers throughout this country experiment with every year. All photography depends upon the camera obscura, as cameras are little more than miniature dark chambers. Under the heading *Photography* the editors of *Encyclopedia Britannica* write, "Discovered by no one man, the outcome of early observations by physicists of the formation of images in the camera obscure and by chemists in the action of light..." These early physicists were ad-Din and al-Haytham, although they are not mentioned so. What the Britannica does note is that the camera was indeed "an outcome of the old camera obscura, originally a darkened room with a tiny hole in the window shutter for viewing natural scenes." Furthermore, the editors write that "the principle had been known to Roger Bacon, Alhazen and others...Leonardo da Vinci," clearly indicating that it was a hotch-potch discovery. Yet, Durant makes it evident that al-Haytham wrote "the first known mention of it" and further notes that, without al-Haytham's works to rely upon, "Roger Bacon might never been heard of."13

Ibn Sina (d. 1037), the great Islamic physician, was also a brilliant physicist. Like many modern students he mastered the science at an early age. Today, physics is a required course for all those wishing to enter the medical field. It is one of the most complex and difficult pre-medical courses. Students have no choice but to study it; most fear it, and many hate it. Ibn Sina's motives were different. He had no need to study physics to pass exams, achieve a degree, or gain licensure. Rather, he did so purely out of interest; he mastered it out of love and curiosity. Thus, he was an original thinker and made numerous innovative discoveries. Perhaps his most astounding statement regards the physics of light. He claimed that light originates from the dissemination of particles from the light source itself (today, these are known as photons), thereby concluding that the speed of light is finite.14 Indeed, it is utterly astounding that this Einsteinian statement occurred in such an early era, yet, unfortunately, Ibn Sina's research is entirely omitted from today's history books.

Ibn Sina performed numerous original experiments in physics, including studies on motion, force, light, heat, and specific gravity. He determined the effects of gravity in a vacuum six centuries before Galileo, which places him as one of the originators of the laws of gravity. While Galileo was primarily a student of physics and astronomy, incredibly, this l0th/11th century genius from Afghanistan became a master in virtually every basic field of knowledge. He was an astute physicist, astronomer, mathematician, engineer, physician, botanist, pharmacologist, zoologist, geologist, philosopher, chemist, historian, and theologian. Astonishingly, his literary achievements were greater than those of Galileo, Newton, da Vinci, Copernicus, Kepler, Boyle, and Lavoisier combined.

A discovery of similar import to Ibn Sina's light theory was made in the 10th century by the renowned philosopher al-Farabi, who gave the first scientific explanation for the speed and cause of sound. Al-Biruni also made accurate measurements of the speed of sound and attempted to determine the speed of light. Although his attempt in the latter fell short, he found it to be much greater than the speed of sound. Thus, he relied upon carefully constructed experiments in attempting to prove his theories. This analytical method was typical of the Muslims of the Islamic Era and gives al-Biruni, Ibn Sina, al-Farabi, Kamal ad-Din, al-Khazini, and al-Haytham the status of being the world's first true physicists.

Astronomy

All prominent civilizations have shown an interest in the science of the heavens. In Islamic civilization astronomy was the first science to attract the curiosity of the scholarly. However, they were not the only ones; the Caliph's were fervent devotees.

During the pinnacle of Islamic rule astronomy became a precise science, far exceeding the level of accomplishment achieved by previous civilizations. "Muslim astronomers understood," says Thomas Welty, "that the earth rotates on its axis and *revolves around the sun* (italics mine)." Furthermore,

Durant writes that the Muslims took the earth's sphericity "for granted." In other words, hundreds of years before Copernicus the modern theory of the solar system was routine knowledge among Islamic astronomers. This was at a time when the Medieval West regarded the earth as a motionless (and flat) object lying at the center of the solar system.

There has been an obsession since ancient times to comprehend the astronomical truths of the heavens. However, the Muslim rulers and scientists had a decisive advantage over all others: the Qur'an. This book makes numerous references to the nature of the heavens and earth, all of which are scientifically accurate. Says the Qur'an, "the heavens and the earth were at first one mass: then, We caused it to explode, and (ultimately) created every living thing out of water (i.e. hydrogen and oxygen)." Students in astronomy classes throughout the world study the Big Bang theory of the origin of the universe. This Qur'anic statement indicates that the theory may be correct.

Concerning the planets the Qur'an says, "each follows its own deliberate orbit." About the stars and galaxies, "God is the one who created the heavens without any supports that you could visualize." About the paths of the planets, "We have created above you numerous celestial orbits." About the endless depths of interstellar space, "Praise is due to God, who has created the heavens and the earth and brought into being profound darkness as well as light." Regarding the pulsating expanding nature of the universe, "It is We who have built the universe with our creative power, and, indeed, it is We who are steadily expanding it." Concerning the orbits of the sun and the moon, "they run their appointed courses." The Arabic word which describes this orbital motion is literally defined as "swim." The existence of the planets of our solar system is implied by "We have created several celestial orbits above you." Regarding the navigational usages of the heavens, "He it is Who has established the stars so that you might be guided by them in the midst of the darkness of land and sea..." Concerning the anciently recognized star patterns, such as the Big Dipper, "Indeed, We have established in the heavens magnificent constellations and endowed them with beauty for all to behold..." In respect to the existence of other solar systems and/or galaxies besides ours, "Praise be to God, Lord of the *worlds*."

It is easy to comprehend how the Qur'an provided Muslim astronomers with an insight of the universe unavailable to previous civilizations. Every effort was put forth by the Arabic-speaking scientists to determine the nature of the heavens. As this was done scientifically numerous original discoveries were made. Thus, it is readily evident that the notion of the sphericity of the earth, as well as the infinite nature of the universe, was brought to Europe exclusively by the Muslims.

Historians credit the Islamic scholars of the Middle Ages with transforming astronomy from a crude, mystical doctrine into a true science. World Book Encyclopedia, among others, implies differently, that is that astronomy remained a cult of the ancients until the advent of Johannes Kepler in the 17th century. This is erroneous, as numerous prominent scientific historians elucidate.[15-19] The Muslims developed astronomy into a scientifically correct discipline over a period of many centuries. They had an early start, largely as a result of the astronomical truths contained in the Qur'an. For instance, their knowledge that the earth is a sphere is Qur'anic in origin. The writings of the ancients were also utilized, but they were not relied upon entirely.

The earth's roundness was known to the Muslims as early as the seventh century, well before they assimilated works by the Greeks on this matter. They were led to this belief through various passages in the Qur'an, including those previously mentioned. One of the words used in the Qur'an to describe the earth's surface, *dahaha*, indicates an ovular shape, that is egg-like, rather than a perfect circle. Such passages were an immense stimulus for the study of astronomy, and, because of the Qur'anic tie to this science, money from the state treasury was generously granted for it. The Muslim astronomers also were well aware that the moon is a sphere orbiting the earth. They also knew that the earth orbits the sun. Furthermore, although they may have been influenced by the Greeks in this matter, they brought forward the fact that the earth courses in an elliptical orbit rather than a circular one. They measured precisely the farthest distance between the earth and the moon and made numerous other studies of the earth's relationship with it.

In 830 A.D. al-Mamum, the Caliph of Baghdad, commissioned a staff of astronomers to test the findings of Ptolemy, ancient Greece's greatest astronomer. Using an innovative technique these scientists measured a terrestrial degree at 56.6 miles, only one-half mile off the most recent calculations. This lead them to make a relatively accurate measurement of the earth's circumference. While their measurement fell short, it was certain evidence that they knew the earth was round and that they were continually attempting to update the science of astronomy through scientific experiments at an early stage in their advancement. Incredibly, Eratosthenes, about 200 B.C., using what were probably less sophisticated instruments, arrived at a highly accurate determination of the earth's circumference of 25,000 miles. This was an astounding achievement in so early an era. However, the Muslim scholars, in their unique style, had to "prove everything" they found in the ancient writings through experiments.[20] Later Muslim astronomers calculated the circumference at 26,000 miles.

One of al-Mamun's staff members, al-Farghani, published in 830 A.D. the text which served as the top authority for astronomy in Europe and Western Asia for nearly 700 years. As a result, Islam had a major influence upon the development of European astronomy during the 15th through 18th centuries.

Islamic astronomers were entirely responsible for renovating European astronomy, dislodging it from astrology and paganism into true science. The fact that the observatory in Mosul, Iraq, had attached to it a library containing 400,000 books is just one example of the research and scholarship that the Muslims devoted to this science. This was a greater number of books than existed in all of Western Europe. This degree of research and documentation remained unmatched until modern times, in fact, many of their achievements have yet to be exceeded. How, then, can Isaac Asimov, with good conscience, claim that the first "scientific" observatories were built in 16th-17th century Europe? He says this despite the fact that the works of Islamic astronomers at the Maragha Institute during the 13th century had a decisive influence upon virtually all early European astronomers, including Copernicus, Brahe, and Kepler.21 The fact that the 16th-17th century European observatories, as well as their instruments, were modeled after Islamic observatories further illustrates the extent of Asimov's blunder.

While the observatories of Europe appeared to be a continuation of the method of Islam, unfortunately, they lacked the intense scholarship and scientific interchange that made Islamic observatories, such as those in Baghdad, Maragha, Cairo, and Samarkand, internationally renowned.[2] Jealousy and pride of discovery, so evident in the likes of Kepler, Brahe, and Newton, impeded scholastic communion. Regarding their libraries they contained a few hundred books at conservative estimates compared to the tens of thousands found in the observatories of Islam.

---

[2] According to Thomas Goldstein the one in Samarkand still stands and may be inspected by individuals visiting the region. Samarkand is located in the former Soviet Union and is now represented by the city/state of Uzbekistan.

Hundreds of Years Before Copernicus...

"Copernicus and Galileo," so every pupil in the Western world is taught, "are the founding fathers of modern astronomy." It is also taught that prior to their advent the only significant astronomical discoveries and/or theories were those of the ancient Greeks, notably Ptolemy. Furthermore, it is implied that the works of Ptolemy and other Greek astronomers were the primary stimulus for the Europeans' discoveries.

These instructors note that Ptolemy, while laying the bedrock for future advances, made numerous severe errors. However, no mention is made of precisely what caused Copernicus and Galileo to make that giant leap from archaic thinking to their timely and scientifically correct determinations. As numerous historians make evident, Europeans were not inclined to radically question the "authority of the ancients."

Will Durant has provided some insight regarding this issue. In his huge book on Middle Age history, *Age of Faith*, he mentions the writings of two Spanish Muslim astronomers as "paving the way" for Copernicus by "destructively criticizing the theory...through which Ptolemy sought to explain the paths and motions of the stars." What's more, according to the *Dictionary of Scientific Biography* the works of Copernicus appear to be essentially a duplication of the findings of Islamic Persia's at-Tusi (13th century). Thus, the missing link for the revolution in European astronomical thought has been discovered. It was the dissemination into Europe of these and other precise Islamic works on astronomy, which were reproduced in Latin from the 12th through 15th centuries.

Few people realize that Copernicus himself maintained many archaic theories regarding astronomy. He held that the orbits of the planets are circular, that the universe is finite, and that the stars are fixed (motionless) to a distant background. Thus, with the exception of placing the sun at the center of the solar system he kept largely with the theories of Ptolemy.

It was Bruno (d. 1600), not Copernicus, who must be credited with being the first European to realize the infinite nature of the heavens, although Muslim astronomers propounded this concept 700 years prior. Said Bruno;

"If therefore the Universe is infinite, why place the earth at its center? The sun, the father of life, is the center of our world; but the center of the infinite universe is in all things...for the center of the universe is neither the sun nor in the sun, neither the earth nor in the earth, nor in any place whatever....Therefore, there are as many centers as there are worlds. . ."

Profoundly, he also believed that there were other solar systems with life forms in the universe. For his views he was convicted in the Inquisition courts of treason and, after serving several years of imprisonment in the dungeons, was burned alive at the stake.

Being the first to point out the inadequacies of the Ptolemaic system for planetary motion, the Muslims paved the way for truth in this science. Among other errors Ptolemy (2nd century B.C.) held that the earth was the center of the solar system and the paths of the planets around it were perfect circles rather than elliptical. He also proposed that the stars were stationary and were attached to a celestial sphere. Ja'far as-Sadiq of Islamic Persia rejected the Grecian suggestion that the earth was the center of the solar system. Additionally, he promulgated the concept of the earth's rotating about itself hundreds of years before such a notion was even dreamed of in the West. It should be no surprise that when in the 16th century Copernicus boldly announced the heliocentric theory he was quick to note his indebtedness to Arabic scholars.

During the 9th and 10th centuries astronomical observatories were constructed in all of the centers of Muslim learning. The fact is the observatory as a scientific institution is entirely an Islamic invention.[22,23]

A primary reason for these extensive developments was that in addition to the devotion of scholars to this science the rulers were intrigued by it. The Caliphs were generous with state funds for financing attempts to further astronomical study. Hundreds of astronomers from all over the world gathered in the Islamic institutions to investigate the heavens. The scholars were paid handsomely, and the observatories were beautifully adorned with sophisticated instruments.

Within these observatories ancient theories of the Egyptians and Greeks were revised. Thousands of errors made by the ancients were rectified and tables were corrected. At the Baghdad school the movement of the sun's apogee was discovered, as was the obliquity of the ecliptic and its progressive diminution. The apogee is the furthest distance of the earth's orbit from the moon, but this term can be applied to its furthest distance from the sun as well. The obliquity of the ecliptic is defined as the angle between the planes of a planet's equator, in this case, the earth, and its orbit around the sun. Currently, this is c. 23 degrees, 26.5 feet. Over 1000 years ago Muslim astronomers measured this at figures minimally different than the modern measurement. In these observatories the distance between the earth and the moon, as well as the earth and the sun, was measured. Sun spots and eclipses were studied. Stars, stellar masses, and planetary motions were analyzed. The positions of stars were systematically mapped and charted; improvements were made upon the Greek efforts. The appearance of comets and other unusual celestial phenomena was evaluated and observations were recorded. Nebulae were discovered.

Of perhaps greater importance was the fact that numerous Muslim astronomers documented the elliptical nature of planetary orbits. This finding revolutionized Western astronomy once it was established. For instance, az-Zarqali of Islamic Spain described the orbit of mercury using the Arabic word *"baydi,"* which means oval.24 For nearly 400 years this finding has been attributed entirely to the German Johannes Kepler (17th century). Thus, it appears certain that Kepler is not the sole discoverer of the elliptical orbit, as all modern textbooks claim. Furthermore, it is likely that he first received this idea not by independent research but instead by reading about it in Latin translations of Islamic books. If this conclusion is definitively proven, the course of history will be drastically altered. The way history is taught on all levels—from grade school to college—will be dramatically changed. Books will have to be revised. Indeed, such a finding would transform the way scientific history is taught globally.

Al-Battani (d. 929) was perhaps the greatest of all early Islamic astronomers. Working in an observatory in what is now a part of Iraq he prepared a highly accurate book of astronomical tables. To aid in the determination of precise calculations he constructed a variety of innovative astronomical instruments, including a sundial, novel types of amillary spheres, and a massive quadrant. The latter device came to be known in Europe centuries later as a triquetem.

As perhaps his most monumental contribution to astronomy al-Battani greatly improved upon Ptolemy's theories of planetary motion. According to C. Ronan his criticisms and corrections of Ptolemy's archaic concepts were "very valuable" in the advancement of Western astronomy. Further, Ronan states that al-Battani's discoveries amounted to "vital points for the future of precision astronomy."25 Says Durant regarding the quality of al-Battani's work, "his astronomical observations, continued for *41 years,* were remarkable for their range and accuracy; he determined many astronomical coefficients with *remarkable approximation to modern calculations...(italics mine).*"

Al-Battani was among the first astronomers to clearly

document the fact that the orbits of the planets changed in diameter, that is they are not perfect circles. Further, he developed incredibly precise methods for calculating the motions and orbits of planets and, thus, played a major role in renovating European astronomy.

Al-Battani's findings heavily influenced Johannes Kepler in his attempts to state the proper rules of planetary motion. In fact, all of the pioneers of astronomy in the West—Copernicus, Kepler, Brahe, and Galileo—quoted the great al-Battani.

According to Ronan al-Battani's works were continually quoted by prominent European astronomers as late as the mid-18th century. Yet, few if any modern history books mention his monumental contributions, accomplishments which were achieved in the 9th and 10th centuries A.D., *700 years before Galileo.*

The illustrious Ibn Yunus of Cairo, the inventor of the pendulum, was yet another precision astronomer of the Islamic Era. During the 10th century he upgraded the entire science of astronomy. In a monumental effort Ibn Yunus edited the great Hakemite Astronomical Table during the 10th and 11th centuries, the accuracy of which surpassed any other. Throughout the entire world it replaced Ptolemy's Almagest and was found in use in distant lands as far away as China. His immense treatise is said to have been over 80 chapters long.26 Even today such a feat would be difficult to duplicate and would require the efforts of teams of modern astronomers.

In the 12th century Muslim astronomers set out to reform the yearly calendar. Their concluded work was highly precise and is more accurate than the calendar that is in use today.

The pinnacle of Islamic astronomy occurred in the fifteenth century. A work by Ulug Beg of Mongul India, a century earlier than Kepler, linked the astronomy of the ancients with that of modern era. Beg, the grandson of the Mongul conqueror Tamerlane, is responsible for the construction of the Samarkand Observatory. Ronan notes that this observatory contained a sextant with a diameter is over 130 feet, making it the largest instrument of its kind in the world. The margin of error of this instrument is incredibly small and is technically described as 4 arc seconds. This is a stupendous achievement, considering that 4 arc seconds is equal to the width of a lead pencil 1 mile away. This incredible precision aided in the calculation of astronomical tables which C. Ronan says, "bear comparison with similar tables today." Thus, Ulug Beg's works are a masterpiece, which if studied by modern astronomers would be regarded as contemporary.

The Muslims greatly improved upon the Ptolemaic system of planetary motion. In fact, they systematically upgraded all Grecian astronomical measurements. Their studies were initially based upon Ptolemy's monumental work, the *Almagest*. The book itself was renamed by the Muslims from the Arabic, *al-majisti*, meaning "the greatest." Muslim astronomers undertook the difficult task of refining Ptolemy's theories and discoveries into a true science. They did so by developing ingenious and increasingly efficient methods of computation. In other words, they joined the exact science of mathematics to the inexact science of ancient astronomy.

The Muslims mastered the art and science of the production of astronomical instruments and were the first to mass produce them: astrolabes, quadrants, sextants, and armillaries of various types. Astrolabes were used to determine the altitude of stars, and, while the original models were made by the Greeks, the Muslims improved upon their design greatly. The perfected models designed by the Muslims were used exclusively by Western astronomers for hundreds of years. Copernicus' astrolabe was of Islamic design. Perhaps most amazing was the astrolabe of Ibn Yunus. Made of solid copper, it was over a meter in diameter.

Amillaries, which are large devices used to map the positions of celestial bodies, as well as astrolabes, were known to the Greeks. However, the Muslims were the first to produce spherical astrolabes. Quadrants were also entirely their inventions. These were used to determine the angular distance of celestial bodies from the horizon. They also created the sextant, a device which measured the altitude of heavenly bodies as well as longitude and latitude. Proof of their discovery is found in the fact that the arm of the sextant is known currently as the alidade, a direct derivative of the Arabic, *al-'idadah* (i.e. ruler). One such sextant had a radius of sixty feet, giving it a higher degree of accuracy than smaller devices.

Muslim astronomers named hundreds of prominent stars, and the English derivatives of many of these remain in the current vocabulary. In fact, the majority of the prominent stars visible in the midnight skies bear names which are derived from Arabic. That is because the star charts utilized by European astronomers during the period of the revival of science (14th through 18th centuries) were created by Muslim astronomers. For instance, the Muslim-named Alcor and Mizar (from Ar. *Mi'zar*) form a visible double star in the center of the handle of the Big Dipper (see Appendix A, Table Two for other names). In fact, the star charts which Europe's famous astronomers and explorers utilized to make their monumental discoveries were Islamic in origin.

Many of the astronomical instruments were also crucial for navigation, whether by land or by sea, particularly at night. Explorers, such as Columbus, Magellan, Diaz, and Vasco da Gama, used Islamic astrolabes, as well as sextants and quadrants, for guiding their way by stellar positions. The precision and workmanship of the Islamic instruments would astound the engineers and machinists of our current era, as they were not only accurate scientific instruments but were also "genuine works of art."[27,28] Because of their efforts in producing highly technical, accurate astronomical and navigational instruments, many of which were exclusively their discoveries, the Muslims were directly involved in the discovery of North America.

The manufacture and/or use of high quality astronomical

instruments was unknown in Medieval Europe. Such instruments were first introduced to Europe by the Muslims in the 10th century. Thus, it is no surprise that the descriptions of instruments constructed by Muslim astronomers in the 13th century indicate a striking similarity to those used by the European astronomical pioneers, the so-called founders of astronomy of the 16th and 17th centuries. The fact is those utilized by Tycho Brahe were essentially copies of the ones constructed at the Maragha observatory by al-'Urdi some 300 years prior.29-31

Perhaps the grandest accomplishment in Islamic astronomy was achieved by a man known primarily for his work in the science of algebra (and his poetry). Omar Khayyam of Islamic Persia gave his country a staggering scientific achievement beyond Galilean thought. Using highly complex mathematics he established a calendar which allows the calculation of the Persian new year not to just the day but on the exact hour, minute, and second that the earth finishes its orbit to start its next solar revolution. While the Gregorian calendar, the one currently utilized throughout the world, has an error of one day per 3,300 years the one calculated by Omar Khayyam errs one day in 5,000 years. Thus, Khayyam's calendar is superior to the modern one.

During the 11th century the scholarly al-Biruni made scientific history by being one of the first men to question the accuracy of Ptolemy's laws for planetary motion. He clearly knew, 600 years prior to Galileo, that the earth rotates about its axis. In contrast, Ptolemy believed it was motionless. Al-Biruni also propounded the astronomical milestone that various climactic and astronomical phenomena can be explained by the earth rotating daily about its axis and yearly around the sun. Thus, he completely debunked the Grecian concept that the earth is the center of the solar system.32 Al-Biruni was so fully aware of the orbital nature of the earth that he, for the first time in history, produced a scientific explanation for the existence of lands "where the sun never sets," i.e. the

North and South Poles.33 In a feat unmatched even in modern times al-Biruni wrote over 60 works on mathematical astronomy, including several books over 500 pages long.34

Also in the 11th century al-Haytham outright rejected Ptolemy's concepts of planetary motion as false. Two centuries afterward at the Maragha institute at-Tusi and colleagues produced models of planetary motion which were consistent with the laws of physics. Astronomy students may recall being taught the *Tusi Couple*, a model of planetary motion which was named after him. Recently, historians have noted a striking similarity between these earlier models of at-Tusi's and the ones advanced by Copernicus, the latter being routinely regarded as the discoverer of the correct theory of planetary motion. Yet, the discoveries of at-Tusi, one of the greatest mathematical astronomers of all time and the head of the internationally renowned Maragha Observatory, directly influenced Copernicus, who was primarily an amateur astronomer.35 Furthermore, it was al-Haytham's astronomical treatise, *On the Configurations of the World*, which clearly had a decisive impact upon Western astronomers, including Copernicus, in their attempts to explain the correct laws of planetary motion. The fact is all that Copernicus wrote regarding the nature of the solar system had already been written in the works of al-Haytham and at-Tusi.36

Numerous other Islamic astronomers, such as Ibn ash-Shatir, as-Sufi, al-Bitruji, az-Zarqali, and Ja'far as-Sadiq, disputed the archaic concepts of planetary motion of the ancients. Using more refined instruments and higher mathematics they provided for the first time in history definite evidence that the planets move around the sun.37,38 Furthermore, they were aware that this movement was not in perfect circles. Thus, they initiated the discoveries of Brahe, Kepler, and Galileo.

Finally, it must be emphasized that it was the Muslims and the Muslims alone who were responsible for the dissemination of the roundness theory for the earth. This principle was largely transmitted to Europe through Islamic Spain, where it was taught in the Muslim universities. Had it not been for this the European explorers would have never dared venture towards the New World for fear of falling off or, as anyone who has seen the reproductions of the maps of Medieval Christendom would understand, the demons they might have encountered along the way.

References

(see Bibliography for more detailed information)

1. Al-Akkad, A. M. *The Arab's Impact on European Civilization.*
2. Bammate, H. *Muslim Contribution to Civilization.*
3. *Dictionary of Scientific Biography.*
4. Nasr, S. H. *Islamic Science.*
5. *Dictionary of Scientific Biography.*
6. Bammate, H.
7. Hayes, J. R. (ed). *The Genius of Arab Civilization.*
8. Corben, H. C. 1991. *The Struggle to Understand: a History of Human Wonderment and Discovery.*
9. *Dictionary of Scientific Biography.*
10. Durant, Will. *Age of Faith.*
11. Sarton, George. *Introduction to the History of Science.*
12. Bammate, H.
13. Bernal, J. *Science in History.*
14. Durant, Will.
15. Hayes, J. R.
16. Ronan, C. A. *Science: Its History and Development Among the World's Cultures.*
17. Durant, Will.
18. *Dictionary of Scientific Biography.*
19. Ibid.
20. Sarton, George.
21. Durant, Will.
22. *Dictionary of Scientific Biography.*
23. Ronan, C. A.
24. *Dictionary of Scientific Biography.*
25. Mansfield, P. *The Arab World.*
26. Durant, Will.
27. Sarton, George.
28. *Dictionary of Scientific Biography.*
29. Hayes, J. R.
30. Durant, Will.
31. De Vaux, Cara. *The Philosophers of Islam.*
32. *Dictionary of Scientific Biography.*
33. Ibid.
34. Ibid.
35. Nasr, S. H. *Science and Civilization in Islam.*
36. *Dictionary of Scientific Biography.*
37. Ibid.
38. Nasr, S. H.

CHAPTER 3

Medicine and Pharmacology

Dark, dreary, and barbaric may be an accurate description of the state of the health sciences in Medieval Europe. Yet, this period produced the Golden Age of Islamic medicine. "It was in the Muslim world," says Thomas Goldstein in his survey of modern science, "that the West first met with a highly developed system of medical care."[1]

The medical disciplines constitute the most prolific and broad fields of Islamic accomplishment. Indeed, the Muslims were the inventors of scientifically based medicine as we know it today.[2-5]

The science of medicine was non-existent in Europe during the Middle Ages. Even the Renaissance failed to establish it. Furthermore, there is no record of the systematic application of the medical sciences in the West until the early part of the 20th century. The Muslims were the originators of a variety of medical disciplines, including general and internal medicine, orthopedics, psychiatry, pathology, gastroenterology, cardiology, parasitology, dermatology, ophthalmology, pediatrics, gynecology, preventive medicine, public health, medical genetics, dentistry, urology, surgery, anesthesia, obstetrics, pharmacology, toxicology, dietetics, and medical ethics. They were also the first to carefully divide the practice of medicine into distinct specialties.

The achievements of the Muslim physicians and educators in these fields are so advanced that they are mind boggling. They are so vast that volumes of books could be devoted to them. The findings are so unique that doctors of the current era could aid their patients by utilizing them.

Prior to the advent of Islam the practice of medicine in Europe was a crude science, in fact, it was a cult. During the Middle Ages the causes of diseases and their cures were unknown. Medicine was based exclusively upon superstition, mysticism, and folk lore. Even herbal therapy was unknown, except by the witches. It was an age when barbers legally performed surgery. This atmosphere might best be elucidated by an eye witness account:

A Muslim physician traveling in Europe was astounded to see, first-hand, the Western cure for head pain; the skull of the pitiful victim was bore open with a hand drill. The rationale? Severe head pain (headache) was a symptom of evil spirits, which were trapped in the skull. Boring the hole allowed the spirits to escape. The Muslim scholar was appalled at the brutality, having no option but to watch helplessly as the "operation" was performed, and, of course, as the patient subsequently died.

Another example is Europe's view that the Great Plague of the Middle Ages was a curse from God. It was the Muslims who attempted to inform the Europeans that its cause was the result of the poor sanitation and lack of hygiene typical of Middle Age Europe. Emphasizing its contagious nature the Prophet proclaimed, "If a region suffers from Plague, stay away from it."6

The Qur'an is a book of laws, which are primarily spiritual and cultural. The elaboration of science is the least of its goals. Nor can its prophet be regarded primarily as a scientist. However, the unlettered Muhammad was a man of numerous qualities. He was a statesman, legislator, political genius, military strategist, and economist. He was a teacher of ethics and a spiritual guide. Plus, he was a man of science.

The Prophet had no schooling; he couldn't read or write. Thus, he wasn't a typical scientist, conducting experiments in a laboratory. He was instead a *perceptive scientist*. In other words,

Muhammad had that innate and rare capacity for comprehending the science behind things without going through the elaborate proving of theorems. His greatest scientific advancements came in the field of human health, of which he was a master innovator. Muhammad maintained a great love for health and hygiene. He advocated personal hygiene and public sanitation as a primary method for preventing the origination and/or spread of disease. He condemned the then common practice of urinating or defecating near potential drinking water supplies. He chastised those who spit in public places. He initiated the science of careful hand washing—before eating, after eating, and after answering the call of nature. He was the first individual to prescribe the rinsing of urine from the private parts. He mandated that the fingernails be kept clipped. Potentially billions of germs may be housed under the fingernails, including the eggs of parasites. As a result of his work hand washing become mandatory in societies in which bathing of any sort was unknown. In a region were water was a precious commodity and bathing a rare accomplishment Muhammad instituted daily washing and weekly bathing, reducing the risks for communicable diseases to nil. The wise Muhammad installed a complete system for public hygiene in a society unaware of the importance of cleanliness, positively affecting the health of millions. What's more, his influence is global and permanent.

A proponent of oral hygiene, he outlined methods for preventing dental disease. He convinced the populous that keeping the gums and teeth healthy was crucial for the maintenance of optimal health.

While having no formal training in nutrition, his ability to perceive the hidden health quality of foods was astounding. He promoted the benefits of pumpkin, squash, cucumbers, honey, yogurt, pure milk, melons, dates, and figs, all of which are nutritionally rich. He was a great proponent of herbs and spices used in food and as medicines. He regarded mineral water as a healthful formulation. The more dense it was in minerals, the more healthy he regarded it. A most interesting revelation was his recommendation for the necessity of salt. Medical authorities today might dispute such an endorsement. Yet, in ancient times it was so highly valued that wars were fought to gain control of the salt trade. Now it is known that salt, as sodium and chloride, is essential to the body and that without it a variety of disturbances occur. Furthermore, salt provides iodine, an element often lacking in food, especially that grown in inland regions. Without a steady supply of this element the body readily degenerates.

Muhammad also recommended vinegar as a favorite condiment, and he consumed it whenever he could. Modern science has proven that vinegar is an excellent source of minerals, especially potassium. He enjoyed and commended the addition of aromatic herbs and spices to foods.

Muhammad was the first individual to recognize the danger of eating carrion, that is animals which have died from unknown or "natural" causes. He warned against the consumption of blood from animals, a common practice of the time. Blood transmits a variety of disease causing organisms. He prohibited the consumption of the flesh of pork and other germ-infested animals of the carnivorous family.

There was no fanaticism in Muhammad's nutrition. Although he usually followed a simple diet, he ate from all food categories: fresh fruits and vegetables, fresh meats, milk products, eggs, and poultry. Because of his recommendations for a varied diet he may be regarded as the founder of the balanced diet. One of his most profound comments was, "The blessing is in the bottom of the bowl." It was in the 20th century that modern science discovered what could be in these juices of Muhammad's recommendations: the vitamins and minerals. Researchers have discovered that as much as 90% of the vitamin/mineral content of food is leached into the cooking medium: the juices, sauces, and water. That is certainly one implication of the blessing of which Muhammad spoke. Thus, Muhammad was the first to allude to the existence of micronutrients, the vitamins, minerals, and flavonoids, within food.

In the field of medicine Muhammad was a naturalist, prescribing various herbs and foods. Honey was his favorite medicinal aid. He prescribed it for a wide range of conditions, but his most prolific recommendation was in the treatment of diarrhea.7 Invariably, his prescription worked. Honey is effective for this condition, as it increases the osmotic pull of the blood, causing the retention of fluids and electrolytes. It also helps speed the healing of the intestinal membranes. Additionally, raw honey contains a number of antibiotic-like compounds, such as phenolic acids and hydrogen peroxide, which effectively inhibit the growth of noxious microbes. Recently, authors writing in the *Journal of Pediatrics* determined that honey is more effective in curing childhood diarrheal diseases than sugar water. The *British Journal of Surgery* describes how raw honey packed into wounds is far more effective than synthetic antibiotics in eradicating infection.

However, the recommendation of fasting is perhaps Muhammad's most dominating concept for superior health. He proclaimed that regular fasting could potentially cure any disease. Muhammad lived this way himself by fasting frequently and implemented the institution of fasting within a society unaccustomed to inhibiting passions. Yet, the greatest reason for fasting is physiological rather than merely self-denial. It appears that excessive food consumption is a major cause of ill health and predisposes individuals to a variety of diseases, particularly cancer, heart disease, obesity, and diabetes. What's more, recent research indicates a modest reduction of caloric intake, as little as 20%, dramatically increases life span. The Islamic institution of controlled fasting, i.e. the month of Ramadhan, is perhaps Muhammad's greatest contribution to preventive medicine.

Thus, the Muslim physicians of the Islamic Era had every reason to pursue the use of natural substances in the treatment of disease: their Prophet prescribed it, and their holy book endorsed it. They accumulated knowledge in the field of natural pharmacology with a passion which was never before seen in the world and has yet to be duplicated since. However, they also developed amazingly high levels of expertise in the fields of surgery, anesthesia, and traditional pharmacology. H. Bammate in his book, *Muslim Contribution to Civilization*, makes it clear as to the origin of Western medicine, stating "Muslim doctors played a decisive role."[8] According to the modern scholar of Islamic science, Emile Savage-Smith, writing in *Western Medicine: an Illustrated History*, the Muslims "produced a vast medical literature," directly influencing the creation of Western medicine. The fact is the basic components of modern medicine were invented by the Muslims. They were the first to turn medical diagnosis, drug therapy, and surgery into sciences.

The Muslims invented pharmacology and its allied sciences. Toxicology, the branch of pharmacology specifically involving the diagnosis and treatment of poisoning, is also their innovation. For instance, Ibn Sina, Islam's most famous physician, developed effective antidotes for snake bites. Regarding the latter he recommended the juice of raw garlic, which is successfully used today. Ar-Razi formulated a variety of emetics (inducers of vomiting) as well as purgatives (laxatives) for the treatment of poisoning. Muslim physicians also invented pharmacognosy, that is the science of using natural compounds from either plants or animals as drugs.

The West was first introduced to scientific medicine through direct contact with the Islamic Empire. According to Goldstein Europeans initially encountered hospitals in Islamic cities such as Baghdad, Damascus, Cairo, and Cordova. Gradually, through imitation of its structure and the translation of Islamic books,

Europe adopted the medical sciences of Islam. In fact, the writings of prominent Muslim physicians formed the basis of the curriculum at European universities. During the 14th century these texts constituted the entire medical library of the School of Medicine in Paris, which was then a tiny facility with a few dozen students maximum. Eventually, this school was renowned throughout Europe. For 600 years, a period over twice as long as the existence of the United States, these masterly texts served as the basis of all medical studies in French, German, Danish, and Italian universities. Thus, it was the writings of Muslim physicians, not those of the Greeks, which formed the basis of medical education during the Renaissance. Ibn Sina's most voluminous work, *Canon of Medicine*, was reproduced as fifteen Latin editions during the 15th century and became the Medieval world's best selling medical book. Interestingly, the works of ancient Greek doctors were far less popular. In addition, it was translated into Hebrew and had a major impact upon Jewish medical practice. The curriculum of medical schools throughout Europe, including those in the British Isles, were based upon the Canon for centuries. It was continually reprinted until the 18th century by which time some 50 Latin editions had been produced. In fact, lectures were given on the medicine of Ibn Sina well into the 19th century. Osler says that his influence is still felt even in the texts written today.

As Ibn Sina's writings formed the basis of Western medical education for centuries, his influence upon 20th century medicine must be registered as vast. Many of his descriptions of diseases may be found essentially unaltered in today's medical textbooks. Certain scholars have written that Ibn Sina's was the most potent influence upon Western medicine of any single person, including Hippocrates. Ibn Sina also wrote a booklet, *Remedies for the Heart*, that contains therapies which if used today would advance the cause of this field of medicine.[9-11]

Today's physicians leave the chemistry and pharmacology

of medicine to the druggists. In contrast, the majority of Muslim physicians were their own chemists and pharmacists. Not only did they become familiar with these sciences, but they also performed original research in them. Thus, while in pursuit of medicinal agents, they synthesized and extracted a variety of potentially beneficial compounds, many of them entirely their discoveries. The medical use of alcohol as an antiseptic was extensive, and they were the first to refine it specifically for this purpose. Asimov, in his *Chronology of Science and Discovery*, claims that it was not until the 13th century that the Spaniard Arnau de Villanova "obtained reasonably pure alcohol for the first time." This is entirely false. The Muslims perfected the procedure of distillation and purified alcohol 400 years prior. Firm evidence for this is found in the fact that the names for different parts of the distillation apparatus are derivatives of Arabic words: alembic (*al-anbeeq*), the flask or upper part of the distillation vessel, and aludel (*al-uthal*), which is the lower part (i.e. the condenser).12 Additionally, they revolutionized the use of acids and bases in medicine. Many of these compounds, including nitric acid, hydrochloric acid, sulfuric acid, and sodium hydroxide, were also produced by distillation. Islam's most famous physicians, ar-Razi (10th century) and Ibn Sina (11th century), both contributed to this effort. Ibn Sina's pharmacopoeia alone contained some 760 drugs, many of which were manufactured on a large scale via the refinements made by the Muslims in the field of chemistry.

Pharmacology and Pharmacy: Islamic Innovations

Despite the interest of Muslim physicians in chemistry and pharmacy they were not its only proponents. In many respects the pharmacists ruled this science. This is because the physicians of Islam were the first to distinguish pharmacy as a separate science from medicine. The Muslims invented the first true pharmacies, where drugs were dispensed by prescriptions. Islamic civilization was also the first to train and license pharmacists. Pharmacy achieved this enlightened status as early as 800 A.D. In Baghdad alone sixty chemists' shops existed, dispensing drugs by prescription.13,14

The pharmacies were regulated by government inspectors, known as *muhtasib*, who threatened the merchants with severe fines and other penalties if they adulterated drugs.15 Pharmacists were required to display a license and had to pass pre- and postgraduate exams to retain it. During this era the Caliph of Baghdad established the first school of pharmacy.16 Thus, it is no surprise that Georgi Zeidon states that modern pharmacologists are wholly indebted to the early Muslims, who were the first to establish modern licensed pharmacies. What's more, all Medieval pharmacies were modeled after the Islamic ones.

It must be remembered that pharmacies in America began as stores renowned for their soda fountains. Incredibly, it was not until the mid-twentieth century that the industry became fully regulated and withdrew from the soda fountain business. Yet, World Book Encyclopedia fallaciously indicates that the science of pharmacy is entirely a Western innovation. Other evidence pointing to its Islamic origination is found in the Western assimilation of the terms alcohol, alembic (flask), alkali, arsenic, and aldehyde to quote some of the "a's" of Arabic alone.17,18

Virtually all of the drugs dispensed by the Muslim pharmacists were extracts, components, or derivatives of natural substances. They were the first to systematically distill essential oils from plants. Motivated by their search for the medicinal components of plants, they extracted essential oils on a large scale via natural distillation. These oils were prescribed topically, as well as internally, for the treatment of a wide range of illnesses. Thus, they may be regarded as the originators of aromatherapy, that is the science of the medicinal use of essential oils. They produced hundreds of essential oils and introduced this technology to the West. The immense value of essential oils as medicines has only recently been reestablished.

Muslim pharmacists also innovated a number of chemical techniques for extracting, concentrating, and purifying drugs. They invented and/or upgraded a wide range of processes, including sublimation, distillation, filtration, calcination, and crystallization. Extracts of rose petals, violet, orange blossom, lemon peel, and orange peel were produced for the first time. These were dispensed as medicines and were also utilized to flavor drugs. They invented filter paper, which is indispensible in modern pharmacology for the purification of medicinal agents. Furthermore, they were the first to use solvents, such as alcohol, for extracting the active ingredients from herbs.

The pharmacists formulated every possible medium for dispensing medications; tinctures, pills, ointments, elixirs, syrups, inhalants, and suppositories. Regarding the latter they were well aware of the absorption of drugs into the blood from the rectum and vagina. Such a scientific approach to pharmacy was simply unknown in any previous society.[19-23]

The Muslim pharmacists stocked their shelves with beautifully designed colored glass bottles and ceramic jars containing liquid remedies, essential oils, flower waters, essences, dried herbs, spices, and tablets. They powdered fresh dried herbs with mortar and pestle for immediate use. Distillation of specific remedies was often performed on site.

While filling prescriptions was one of their major roles, these pharmacists were not mere dispensers. The entire affair of selecting the appropriate drug was a science itself and was managed by the pharmacist's guide book: his/her pharmacopoeia, a concept still utilized by today's pharmacists. This method was adopted by Europe in its entirety during the post-Renaissance period.

From there it was brought to the United States.

The heightened interest of the Muslims in the use of natural compounds as healing agents was a consequence of the Islamic belief in God as the creator of Nature. God provides for his creatures: He has placed upon this earth medicines of all kinds. In the words of the Prophet, "For each disease, God has created its cure."24 These beliefs were a tremendous motivation for Muslim physicians, pharmacists, and botanists to study and accumulate the remedies of Nature.

The interest in the medicinal value of herbs caused Islam's physicians and pharmacists to become specialists in botany. Ibn Baytar traveled the furthest extent of Europe and Asia to study medicinal plants. His book, *Singular Cures*, offers many treatments still unknown to modern physicians. His greatest work, the *Compendium of Pharmacology*, which was published in the 13th century and which described the medicinal powers of hundreds of herbs, is such a monumental work that it may be rightly called the most influential pharmacopoeia ever. William Osler, who is widely regarded as the father of modern medicine, says that works such as the Compendium caused the Muslims to have a "heavy hand" in influencing the direction of modern pharmacology and in determining the breadth of modern drug lists.25

Ibn Sina was perhaps the most renowned physician-pharmacist of the Islamic Era. As a physician he is regarded as one of the greatest of all times, races, and cultures. He wrote entire libraries on pharmacology and medicine in his rather short lifetime (he lived to be only 57). His pharmacopoeia included a thorough list of Grecian drugs, which he carefully defined and updated plus numerous original findings. He left detailed instructions for each drug regarding how to use it, which conditions it cured, and any side effects it might exhibit. He also gave a variety of methods of preparation.

The accomplishments of the Muslims in pharmacology are greater than any previous civilization and any since, with perhaps the exception of our modern one. However, in the field of natural pharmacology they remain untouched. Even today's pharmaceutical giants have yet to exceed them. The Muslim pharmacologists and botanists made an utterly immense contribution with their careful and detailed descriptions of thousands of herbal medicines. Unfortunately, most of their works have been lost in antiquity. However, much of this knowledge has been preserved in Ibn Baytar's 13th century book. It is a huge compilation, revealing a list of some 1400 drugs, many of which could benefit humankind if utilized today. A large number of these drugs were their own discoveries. Their formularies include detailed descriptions of the drugs' geographical origins, physical properties, therapeutic usages, and the methods of preparation. Here, they implemented the processes of chemistry to isolate and purify the active ingredients, processes which are integral components of the modern drug industry's manufacturing procedures. In closing it was H. Bammate who noted that the Muslims added some 2,000 different natural medicines to Dioscoride's Herbal (1st century A.D.), an accomplishment which fully controlled the future direction of pharmacy well into the 20th century.[26-28]

The importance of Islam's accomplishments in these sciences is underscored by the fact that medical theory and study, as well as the trade of pharmaceutical substances, was banned in Middle Age Europe well into the 15th century. This is in contrast to the environment in 12th century Baghdad, where some nine-hundred doctors from all over the world arrived to take exams under the direction of Islamic professors. Physicians of the entire Islamic Empire from Spain to Baghdad and Afghanistan were required to pass licensure exams in order to practice medicine. In this manner hucksters and charlatans could be screened out.[29-31] It was King Roger II of Sicily who during the 12th century was the first European to mimic this method.

Medical education in the Islamic world during the 9th through 15th centuries was unique as compared to the current method. Muslim scholars saw no need to secularize medicine. Faith in God and high moral character were regarded as assets. The scholarly Ibn Rushd said about medical education, "Whosoever becomes fully familiar with human anatomy and physiology, his faith in God will increase." Ibn Rushd's medical philosophies were so highly regarded in the West that his works were studied, notably in the University of Mexico, as late as the mid-19th century.

While an extinct science in all of Europe during this era, medicine was thriving in the Middle East. In the epitome of intellectual decadence Western Europe described disease as a "punishment from heaven to exact the penalty of sin," while Muslim physicians had already proven that the Bubonic Plague was a contagious disease and nothing more and even offered the mechanism for its treatment. As said Muhammad, "Pray, but take your medicine too." Thus, Islam, once again, kept in perspective faith, reason, and science.

The Hospitals of Islam

The establishment of Muslim hospitals was one of the most dominating advances in medicine during human history. Historians have clearly demonstrated that the first true hospitals were a product of Islamic civilization. Some scholars have commented that many of these institutions would have rivaled, perhaps surpassed, our modern ones.32-35

The hospitals of Islam were renowned throughout the world for the quality of care, excellent results, and aesthetics they provided. Their structures contained magnificent courtyards, beautiful flower gardens, and fountains of pure water. Kitchens, chapels, nurses stations, pharmacies, and floral gardens were also found within them.[36-38] Some were even adorned with date palms, which gently swayed in their courtyards. According to Thomas Goldstein as many as thirty such hospitals were in existence by the 13th century.

Cleanliness was of the utmost importance within the Islamic hospitals, and the Muslims developed an innovative method for recirculating and purifying the air.[39] The water was also purified and recirculated.

The hospitals of Islam were systematic in structure and operation in every phase of their operations. Each was directed by a hospital administrator. Nursing stations were under the command of supervisors, who were in turn responsible to the head nurse. Notes were kept on each patient and were the responsibility of the medical students. Many hospitals had attached lecture halls for the furthering of medical education. Certain of the hospitals housed medical schools.[40,41]

For educational purposes, as well as concerns of patient safety, the hospitals were divided into various wards. Muslim physicians innovated the use of isolation wards for preventing the spread of contagious diseases. There were separate wards for women and special maternity wards to maintain privacy. All of the major hospitals maintained pharmacies and kept large botanical gardens for cultivating medicinal herbs.[42]

In Islam's hospitals the attributes of kindness and mercy were equally as important as medical acumen and cleanliness. As a result, the Muslims were far ahead of the Europeans in the treatment of the insane. Most hospitals maintained a separate wing specifically for lunatics and other mentally ill individuals.[43,44] In fact, Islamic physicians innovated the concept that mental illness is a distinct disease treatable with medications and psychotherapy. During the same period and for many centuries afterward the insane were treated brutally in Europe. In fact, the current treatment of them in the West is appalling.

The hospitals of Islam served people of all races, colors, and religions. Wealth was not a requirement for entry. The poor and indigent were cared for equally, in other words, anyone could get free medical care. This was due in part to the fact that the hospitals were controlled by the government rather than for-profit or clerical/non-profit agencies. Even enemies, that is prisoners of war, were given equivalent treatment.45

The sophistication of Islamic health care seemed to know no limits. Rural health care outposts were established, the first of their type anywhere in the world. Care was also extended to the inner city. The latter was the first model for the current county hospitals found in every major city in the United States. For the first time in history mobile medical units were dispatched to assist in the treatment of those unable to travel to major cities. What's more, records reveal that Muslim physicians made house calls, even to jails.46

Physicians of all faiths worked together within the Islamic hospitals in complete harmony, the only goal being patient welfare and the cure of disease. Interestingly, concern for patient welfare was so pronounced that patients were routinely given five gold-pieces upon discharge so they could rest and not have to return to work immediately.47 Such an approach in patient care is simply unknown in today's medical institutions.

Islamic doctors were conservative in their approach to patient care. They were true followers of Hippocratic oath. Thus, medical disasters were relatively uncommon. The initial treatment within the Islamic hospitals consisted of diet and physiotherapy, which included water baths and exercises. Food was healthy but rich, plus it was elaborately presented. If these had no effects, herbal and drug therapies were implemented. Surgery was used only as a last resort.

The Islamic physicians utilized every means possible in attempting to cure patients. If the patients appeared to be distressed or if they couldn't sleep, musicians arrived to play soft music, or they would be furnished with a professional story-teller to hopefully distract them from their worries. This is in contrast to modern hospitals, where nutrition, physiotherapy, relaxation therapy, and herbal medicine play no significant role. Jello, ham, bacon, white sugar, soda pop, French fries, and white bread, among other nutritionally inadequate foods, are still on the menu. What appalling things to feed the chronically ill. As for relaxation nothing of it is found except, perhaps, a few worn-out novels and jig saw puzzles.

In the fields of cleanliness and hygiene the Muslim hospitals rose to heights bar none. According to Gustave Le Bon Muslim hospitals were superior in this area compared to many of the modern ones. These hospitals were constructed large enough so that air and water circulated freely through them.48,49 This was because the Muslims recognized the critical importance of stagnant air and water in the spread of infectious disease. They also knew the value of a continuous supply of fresh air in promoting good health and in preventing infections. The first procedure for sterilizing air was developed in these hospitals, a method which has to this day eluded modern medicine.50

In contrast, the hospitals in Medieval Europe were disgustingly filthy. Florence Nightingale reported that mold and algae could be found growing on the walls within some of the hospitals. The floors, she said, were covered with debris, organic matter, dried blood, and other filth. Bedding remained unchanged between patients; the mattresses were soiled with blood and excrement. Not surprisingly, few people made it out of them alive. Such gruesome conditions existed well into the late 19th century.51

This issue of attention to antisepsis and hygiene cannot be overemphasized. Infectious disease is the primary cause of illnesses in hospitals today. In fact, currently there is a virtual epidemic of infections which are contracted within the hospitals. These are called in medical terminology *nosocomial infections* and account for tens of millions of illnesses and hundreds of thousands of deaths every year. According to a recent article (1997) in the *Journal of the American Medical Association* as many as 250,000 deaths from hospital acquired infections occur yearly in the United States alone. Billions of dollars in research have been spent in order to arrive at the cause and/or solution of this serious dilemma. The most recent conclusion: inappropriate hygiene, particularly failure to attend to hand washing, is the major factor in the spread of the infections. Ironically, while the Islamic hospitals were massive and housed thousands of patients, they were generally free of such epidemics. To again quote Gustave Le Bon the hospitals of Islam "appear to have been built under conditions which, from the point of view of hygiene, were *greatly superior to our present day establishments* (italics mine). They were enormous and air and water circulated in them quite freely..."52 It is unfortunate that the builders of 20th century hospitals were unaware of these ancient hygienic secrets, because many thousands of lives could be saved every year.

The epitome of this concern during the Islamic Era is exemplified by ar-Razi's method for determining the most hygienic region of Baghdad for its new hospital. *He hung strips of fresh meat* in different districts of the city to see which ones would decay most rapidly. The area where the meat decayed the least was the selected site.

In contrast, it was not until late in the nineteenth century that Halsted and colleagues, after much harassment and persecution, were able to implement a standard "washing of the hands" policy to prepare surgeons and nurses for surgical procedures. For centuries, American and European surgeons regarded it as heresy to cleanse the hands before they operated. As a result, more than 50% of all surgical patients died from rampant infections.

As late as the 19th century the majority of the surgeries occurring in the United States and Europe were performed by non-medical unlicensed individuals such as barbers and shepherds. Thousands died unnecessarily from infections resulting from the procedures themselves and as a result of the contamination of internal organs and wounds by germs carried on the doctors' hands. While for the West the now standard preoperative procedure of washing the arms to the elbows is essentially a twentieth century invention, Muhammad, may God rest his soul, instituted a systematic personal hygiene procedure which included this practice over 1400 years ago. Such cleansing procedures were standard for the surgeons of the Islamic Era. To them, it was heresy for the public to neglect hand washing, let alone surgeons.

In the 12th and 13th centuries the Islamic hospitals, under the patronage of the Caliphs, reached their zenith. The hospitals in Syria and Egypt achieved such heights that travelers and historians regarded them as one of the greatest treasures of civilization. They attracted gifted students of all faiths and backgrounds from throughout the world. Christians, Jews, Zoroastrians, Bhuddists, and Muslims all interned there. The elegant and spacious buildings were well equipped with educational facilities, lecture halls, and libraries. There were pharmaceutical laboratories on site, where medications were freshly prepared and dispensed. For this purpose botanical gardens abounding in hundreds of medicinal herbs were carefully maintained. Numerous historians regard these hospitals as the crowning achievement of Islamic culture, architecture, and civilization. The Islamic hospitals were so magnificent and the care so wonderful that many thousands feigned illness just to enter their gates.

The Cairo hospital was perhaps the most impressive. During the 13th century it had been expanded and was so huge that it contained several thousand beds. Privacy was maintained by the existence of separate wards for males and females. Will Durant aptly describes this lost spectacle: "Within a spacious quadrangular enclosure four buildings rose around a courtyard adorned with arcades and cooled with fountains and brooks. There were separate wards for diverse diseases...and particularly pleasant accommodations for the insane."53 Incredibly, the world famous hospitals of Cairo, Damascus, and Baghdad remained in continuous operation for some 300 years, offering free medical services and dispensing drugs at no charge. To comprehend the immense scope of this achievement this represents a period of time nearly 100 years longer than the existence of the United States (1792 to date). Thus, the Islamic hospitals, with their beauty, grandeur, excellence of care, and utter tolerance of humanity, have never been superceded, whether in modern or ancient times.

Nutrition and Dietetics

The Muslims were the first to recognize the importance of nutritional defects in the cause of disease. They had an early start. Prophet Muhammad was a major proponent of proper diet. In fact, he was the first individual to indicate the existence of vitamins and minerals, although he never named them as such. He was also the first to propound the immense healing properties of certain foods. For instance, he was a major proponent of foods rich in pigments such as dark grapes, squash, watermelon, and pumpkin. Now we know that these pigments, the carotenes and flavonoids, are among the most powerful disease fighting chemicals known. Muhammad specifically recommended mineral rich foods such as the potassium-rich vinegar, coriander, and dark honey. He touted heavy mineral water as a cure-all. Apparently, he was fully aware that a lack of minerals could devastate health.

Muhammad never recommended caustics or potentially poisonous compounds. In contrast to today's physicians, who freely dispense toxic substances, Muhammad only recommended safe natural items. In fact, he warned against doing the body harm. What's more, he regarded excellent health as the greatest of all virtues.

Motivated by Muhammad, Muslim physicians and pharmacists of the Islamic era became the world's original nutritionists. They recommended fresh nutritionally rich food as the foremost therapy. This was supported by herbs, spices, and aromatic waters, all of which aided in the nutritional support. Islamic hospitals emphasized fresh food as a cure, offering it as the initial treatment. An entire system was created for dispensing certain foods, herbs, and spices for treating specific diseases. High protein foods, as well as fresh fruit and vegetables, were a mainstay of treatment. Furthermore, the food was carefully prepared and presented elegantly. Thus, diet became a science under the auspices of Islam.

Ibn Zuhr of Islamic Spain wrote the world's first manual on diet, *The Book of Diet*, in which he described in detail how to treat and prevent disease through nutrition.[54] Islamic scholars were the first to accurately describe the function of the digestive organs. They demonstrated the physiology of the stomach, describing both its secretions and movements 1000 years before Beaumont, to whom this discovery is erroneously attributed.[55] Al-Masihi correctly explained that the absorption of food occurred primarily in the small intestine; prior to him it was believed that absorption occurs in the stomach. Furthermore, Ibn Zuhr propounded the value of nutritional therapy for the debilitated. He fed hospitalized patients through stomach tubes and nutrient enemas 1000 years before this was known in the West.[56] These facts make it evident that the science of dietetics, as well as nutrition, is an Islamic innovation.

The majority of Islamic physicians recommended only natural compounds, avoiding caustic agents/drugs. In other words, following the Prophet's philosophy they made every effort to avoid harming or killing patients. Such a philosophy would be of great value today. Incredibly, as many as 40% of the admissions in some hospitals in the United States are due to drug toxicity.

Obstetrical Care

In obstetrics the use of emergency Cesarean section was perfected by Muslim physicians. They performed it nearly as efficiently and effectively as today's obstetricians and only utilized it after all other methods were exhausted. However, most vaginal deliveries were performed by midwives and obstetrical nurses rather than physicians.[57-59]

The practice of midwifery flourished in the Islamic Era and has failed to reach such a level of excellence since. Women were usually delivered in the squatting position, the most natural and effective means for fetal expulsion.[60] As an additional testimony to the level of excellence achieved in midwifery during this era Islamic Spain's az-Zahrawi in his textbook of medicine devoted an entire chapter to it.

Muslim physicians, nurses, and midwives were fully aware of the importance of maintaining cleanliness and antisepsis during the birthing process as well as in the postnatal period. The doctors and midwives practiced careful hand washing technique prior to performing the necessary procedures. Evidence exists that Muslim surgeons and midwives washed with natural antiseptics, such as rose and orange extracts, prior to deliveries or surgeries. As a result, maternal mortality rate during the Islamic Era was lower than ever before in history. In contrast, in Medieval Europe women were fortunate to survive even relatively normal labors; for these, the mortality rate was nearly 50%. According to F. H. Garrison's account of 16th century Europe in the tougher cases, women were "usually butchered to death."[61] This was at a time when deliveries were attended by knife wielding herdsmen or barbers. Eventually, a law was passed in the latter part of the 16th century prohibiting these knife happy scoundrels from attending obstetrical cases. However, it was not until the 20th century, after many centuries of debauchery, that the high rate of maternal, as well as infant, mortality in the Western world was resolved.

Surgery and Anesthesia

The historian Haidar Bammate claims that of all medical disciplines Islam's most valuable contribution was in the field of surgery. The Muslims of the 10th through 13th centuries were the world's first true surgeons.[62-66]

The leading Muslim surgeon was Islamic Spain's Abu'l Qasim bin 'Abbas az-Zahrawi, who died in 1107 A.D. He was popularly known in the West as *Abulcasis*. His works have been acclaimed as the main source of inspiration for all surgeons from the 12th through 16th centuries, a period of over 500 years.[67-69]

While erroneously held as a modern discovery, anesthesia was almost entirely originated by the Muslims. They utilized a variety of herbal extracts to partially or completely sedate patients prior to performing surgical procedures. They were the first to systematically use inhalants as anesthetic agents. One of their herbal drugs, called *darnel*, was apparently a potent narcotic, as its administration resulted in the patients becoming completely unconscious, perhaps as long as a week.[70] Additionally, Islamic physicians developed an innovative form of nasal anesthesia. They rapidly anesthetized patients by applying narcotic-saturated sponges over the face.[71-73] Thus, they were the first to realize that rapid induction of anesthesia could be accomplished through the respiratory passages.[74,75] Furthermore, they routinely used refined opiates to render patients unconscious, a practice which is instrumental in modern anesthesia. Europeans first learned of this practice directly from translations of Islamic books. Other anesthetic agents used by Islamic anesthetists included hemlock, hyocyamus, belladonna, lettuce seed, mandrake root, ice, marijuana, and aromatic oils, all of which they described in detail.[76-77]

Muslim surgeons developed numerous previously unknown surgical instruments, medical devices, and procedures. They invented the hypodermic needle. This was utilized to perform suction surgery in the removal of cataracts.[78,79] This device, also known as the intravenous needle, is of the utmost importance and is still used today in every medical institution throughout the world. They perfected the procedure of *lithotrity*, which is the crushing of bladder stones prior to their removal. For this purpose they developed a unique method of widening the urinary tract to aid in the diagnosis and treatment of urinary tract diseases. Common use of lithotrity rather than surgical removal of the entire stone has only recently been reactivated.

Urinary catheters were a necessary instrument for the surgeons of the Islamic Era, since they used potent sedatives to induce anesthesia. It was az-Zahrawi of Islamic Spain who brought the use of this critical surgical device to the West. He invented several types of catheters.[80,81]

The Muslim surgeons of Spain were the first to perfect the manufacture of surgical suture from animal intestines, also making it from silk and wool. In fact, the surgical suture in use today may be attributed to them. They mastered the use of cautery, whether by heat or acids, to control bleeding during surgery. Today's surgeons frequently use cautery for this purpose.

Ibn Zuhr (13th century) of Spain was the first to mention the surgical procedure bronchotomy, which is the removal of a section of the lungs. He also gave the first correct surgical description of pericarditis, that is inflammation of the heart sac. This is a highly significant finding, since it illustrates the deliberate effort he undertook to carefully record symptoms and physical signs of specific diseases. His books contain detailed instructions for surgical reduction of fractures.82 His "bone setting" methods were relied upon by Europeans for many centuries after his death, a fact which secures him as one of the pioneers of modern orthopedic surgery. Additionally, Ibn Zuhr was the first to recognize the occurrence of middle ear infections. His determination of the cause of scabies through the identification of its organism, the mite *Sarcoptes scabe*i, establishes him as the second known parasitologist.83

It was Ibn al-Quff (13th century) who gave the most comprehensive description of its kind regarding surgical operations for the treatment of trauma. Yet, in the field of general surgery Abu'l Qasim az-Zahrawi's treatise was most influential. Among his accomplishments were the first clear description of hemophilia as a hereditary disease. He also discovered the use of the so-called Walcher position for obstetrics, which he originated some 700 years before Gustav Adolf Walcher was born.84 With slight modifications, this is the position currently used in obstetrics. Az-Zahrawi also provided the first accurate description of genetic deformities of the mouth and dental arches. He correctly described the pathology behind paralysis of the vocal cords hundreds of years before this was discovered in the West. Furthermore, az-Zahrawi's detailed, accurate descriptions of inherited diseases, many of them his own discoveries, qualifies him as the world's first known medical geneticist.85

Az-Zahrawi performed numerous innovative surgeries, including hook extraction of nasal polyps, speculum removal of bladder stones, and reductions of a variety of dislocations.86,87 This Islamic surgeon was a master in the medical discipline of orthopedics. He essentially invented the use of plaster of Paris for casting and was the first to pad splints for broken bones with cotton.88,89 Interestingly, prior to az-Zahrawi's discovery Muslim physicians applied linen soaked in egg whites against splints as a casting material.90 They eventually perfected casting for setting bones with az-Zahrawi's plaster casts in the 11th century. Impressively, the cast we know of today is essentially their innovation.

Az-Zahrawi described numerous surgical procedures with modern perfection. He ligated (i.e. tied off) blood vessels five centuries before the Frenchman Ambroise Paré, to whom the procedure is wrongly attributed. Descriptions of some 200 surgical instruments are found within his writings, many of which were employed by Christian surgeons for over four hundred years after his death. Rather than the crude devices found in ancient Greece az-Zahrawi's instruments boasted precision workmanship, and some were outright works of art. His ingenious devices included speculums, scalpels, syringes, catheters, and the tongue depressor, the latter being solely his invention.91-94 The fact is az-Zahrawi invented hundreds of surgical instruments, and drawings of his unique devices exist currently.95

The translation of az-Zahrawi's works into Latin sparked a revolution in Western surgery, which was, prior to him, barbaric. His books introduced the science of modern surgery to Europe, in which it was previously unknown.

Az-Zahrawi was unique compared to the surgeons of today's age. He was not of the aptitude to rush into surgery as the answer for a given health problem, nor did he rely on it as the sole method of cure. Rather, he promoted preventive medicine. First, he recommended that patients improve their diets and nutrition. If this was unsuccessful, he prescribed drug or herbal therapy. Surgery was performed only as a last resort.96

Az-Zahrawi maintained a philosophy from which many doctors currently could benefit. He warned fellow physicians and medical students against performing dangerous procedures with the clouded intention of achieving material gain. As a master diagnostician he stressed the importance of careful observation of the patient's presentation in order to secure the most accurate diagnosis. No matter how much time it took to procure, he felt, an accurate diagnosis was the crucial first step for achieving a cure.

The interest of the Islamic physicians in surgery caused them to be original researchers in the disciplines of anatomy and physiology. In the 13th century Ibn an-Nafis' text provided the first detailed and correct description of the pulmonary circulation. Ibn an-Nafis went further to accurately describe the physiology of the heart, i.e. that it is a pump. Incredibly, he also gave the original correct description of the microcirculation of the lungs. In his treatise he explains how the blood vessels in the lungs gradually become smaller and smaller. In the tiniest vessels, he says, "air" from the lungs is absorbed.[97,98] This was the first correct reference to the oxygen absorbing function of the lungs, which clearly qualifies Ibn an-Nafis as being the world's first known pulmonologist. In another astounding scope of genius Ibn al-Quff accurately described the anatomy and function of the capillaries, even though he had no microscope to see them. This discovery is erroneously attributed by Asimov and others to the Englishman William Harvey of the 17th century.[99,100]

Advances by Muslim anatomists in the 13th century put to rest the variant theories concerning the septum of the heart. The septum is the barrier between the right and left sides of the heart. Since Roman times it had been thought that the septum contained tiny pores, which permitted blood to be passed from one side of the heart to the other. The Muslims challenged the distorted theories of the ancients, proving that no such pores exist. They found instead that blood enters the left side of the heart after it flows through the lungs.[101] Even Vesalius himself, the "father of anatomy," believed in this archaic idea, as did Leonardo da

Vinci.102,103 'Abdul Latif (1161-1231 A.D.) made an additional advance in the science of osteology, that is the study of bones. While studying human skeletons in Egypt, he wrote the first critique of Galen's views of skeletal structure, paving the way for a more accurate study of the bones and joints.104

The illustrious Ibn Zuhr (12th century) was another expert surgeon of Islamic Spain. He was the first to describe the existence of wet and dry pericarditis (inflammation of the sac of the heart), tuberculosis of the internal organs, and mediastinal abscesses (the mediastinum is the cavity located behind and around the heart).105,106 Furthermore, Ibn Zuhr gave a precise explanation of the action of the heart valves, the first of its kind, and how they open only in a single direction in order to keep blood flowing in a one-way channel. These findings were of immense importance for the advancement of modern medicine.

The scope of Ibn Zuhr's genius is manifested by the fact that prior to him no one correctly described the physiology of the heart. In Ibn Zuhr's descriptions there is complete independence from the distorted theories of the ancients, since it was he alone who correctly elucidated the function of the heart as a one way pump. The accuracy of his writings regarding the specific function of the heart and other human organs qualifies Ibn Zuhr as the founder of human physiology. His descriptions of organs and their functions are far more accurate and voluminous than those of William Harvey, to whom the discovery of this science is usually attributed. The fact is Ibn Zuhr's descriptions of various medical conditions are so accurate that little could be added to them today.

Muslim physicians corrected numerous anatomical errors made by the ancients. They studied anatomy specifically for improving surgical technique. They were the first to provide anatomically precise descriptions of the internal organs, particularly the heart, liver, spinal cord, and lungs. These and other findings establish az-Zahrawi, Ibn an-Nafis, Ibn al-Quff, Ibn Zuhr, 'Abdul Latif, and numerous other Muslim physicians as being among the first anatomists.

It is difficult to understand how Galen, even though he was the most famous physician of Roman times, could be regarded as one of the founding fathers of anatomy. He performed most of his surgeries on wounded gladiators, learning anatomy by trial and error. Boorstin notes that his numerous blunders have earned him the title of "the ape of anatomy."[107]

Cautery was utilized extensively by Muslim surgeons and not simply for the control of bleeding. They used it to destroy skin cancers and to open abscesses. Today, it is also used for these purposes.

Few historians have recognized that the tourniquet is an Islamic development. It was al-Majusi of Islamic Persia, known in Latin as *Haly Abbas*, who invented it.[108,109]

Despite these vast achievements in surgery, surgical instrumentation, and surgical technique, Western textbooks give the impression that the practice of surgery was neglected in Islamic medicine. The majority of these texts suggest that the Muslims never seriously contributed to surgery as a science. However, the works of az-Zahrawi alone thoroughly dispute this; there were thousands of other Islamic surgeons. It was az-Zahrawi who was the first surgeon to provide specific protocols for the study of anatomy. He wrote:

> Before practicing one should be familiar with the science of anatomy and the functions of the organs...to understand them, recognize their shape, understand their connections, and borders. The surgeon should know the bones, nerves, muscles; their origins and insertions, the arteries and veins, from beginning to end.[110]

This is precisely the modern science of anatomy, practiced by az-Zahrawi and colleagues over 1000 years ago. The ubiquitous claim of the existence of an Islamic void in surgery is further debunked by the fact that today's surgeons are utterly dependent upon such surgical devices as the intravenous needle, tourniquet, scalpel, winged tip clamp/scissors, speculum, forceps, and the urinary catheter, all of which were invented and/or perfected by Islamic physicians. It was H. G. Wells in *Outlines of History* who made it clear that Muslim physicians performed "the most difficult" surgeries known, a fact which is clearly illustrated by the refined surgical instruments they used. After all, it was an Islamic book which served as the surgical authority in the West for over 400 years, a feat which has yet to be duplicated.

Ophthalmology

Ophthalmology as a form of deliberate medicine owes its existence to Islamic civilization.111 The achievements of Muslim physicians in this field remained unsurpassed by the West until the 20th century.

Ar-Razi, writing in the 10th century, is thought to be the first to have correctly described the physiology of the pupillary reflex. Ibn Sina's contribution to ophthalmology included the discovery of the total number of extraocular muscles as six and the correct description of their function. 'Ali Ibn 'Isa's text, *Memorandum for Eye Doctors*, remained the authoritative one throughout Europe and the Orient well into the 19th century. Al-Haytham carefully delineated the anatomy and physiology of the eye. He gave the first detailed description of its connection to the brain via the optic nerve. He correctly determined that when light strikes the retina the information is delivered to the brain through this nerve. In other words, he discovered that the optic nerve transmitted impulses to the brain and that this resulted in visual projection.

His findings were disseminated throughout Europe during the Renaissance and were studied in Europe's medical schools for centuries. Spanish Muslims made the first correct explanation of the physiology of the retina; the term itself is derived from Arabic. Hunayn's ten books (9th century) on anatomy and physiology included an extensive section on the anatomy of the eyes, said by S. Hamarneh to have had a profound influence on the development of ophthalmology in Europe.[112] His anatomy charts of the eyes were utilized extensively for several hundred years. Ibn al-Kawafir's 13th century work, *Diseases of the Eyes*, was the most scientific book ever published on ophthalmology and remained unsurpassed in Europe until the 19th century. The influence of his book permeates today's ophthalmology texts.

In the 13th century al-Mahusin invented the hollow (hypodermic) needle and used it to perform the first cataract operation via suction. He became so proficient at it that he successfully performed the operation on a one-eyed man.[113]

Additionally, Muslim physicians and pharmacists developed a number of drug treatments for the eyes. They formulated topical remedies (eye drops) for the treatment of conjunctivitis, eye infections, and traumatic injuries.

Medical Philosophy

The philosophy of the physicians of the Islamic Era was entirely different than that of our modern one. It was clear to them that the human body and the divine existence are inseparable. According to Islam the divine spirit permeates the mind and soul, in fact, the very cells of our bodies are infused with it. Comprehensive care must include treatment of all aspects of the human creature—physical, mental, spiritual, and emotional. Thus, the Islamic approach to patient care and healing is unique. Medicine and prayer were combined under the direct supervision of the physician possibly for the first time in history.

Profit was rarely a motive for the Islamic physicians, nor was it an issue within the hospitals. Funds from the rich Islamic treasury cured that concern. Quality of care was the only priority. Thus, there were no conflicts of interest. Plus, as has been previously stated, they were more than just physicians. They were statesman, philosophers, jurists, authors, scientists, and scholars.

The medical philosophy, as well as ethics, of Islam's physicians had a decisive impact upon Western medicine. The ethics of Islam, that is the use of safe, natural, scientifically based, and conservative approaches to health care, provided the Europeans their first opportunity to abandon their barbaric practices. The medical philosophy of Islamic doctors, such as Ibn Sina, Ibn Zuhr, and Ibn Rushd, was so highly regarded in Medieval Europe that the names Hippocrates and Galen were rarely mentioned in comparison.[114-116] Ibn Zuhr's comment about the natural healing process was prophetic. The crux of his philosophy is that the natural powers of the human body are what ultimately cure most diseases.

Spain's Ibn Rushd was possibly the logician supreme of the Islamic physicians. Born in Cordova in 1126 he came from a long line of distinguished scholars. He exceeded the achievements of all of his forefathers and became an internationally renowned expert in law, medicine, and philosophy. Yet, his was no ordinary philosophy. Says M.A. Martin, Ibn Rushd "specialized in making all things clear." That is why he was more popular during the Middle Ages than Aristotle.[117]

To Ibn Rushd a complete cure could more readily be achieved if the turmoil of the soul was resolved. He said, "I believe the soul is immortal, but I cannot prove it."[118] This was a man who, rather than depending on emotions, dogma, or personal experience to explain the nature of man's existence, relied entirely on reason. Man's purpose was to discover truth, Ibn Rushd insisted, and the serious study of God and His workings constitute the most profound, noble form of worship. It is said that he never missed an evening in his adulthood of writing and reading except on his honeymoon and when his father died.119

Ibn Rushd simplified much of the philosophy of Aristotle and wrote far greater volumes, which contained views entirely his own. His writings, once translated into Latin, gave the Western world its first glimpse of Hellenistic philosophy, although it was a much improved and enhanced version. These writings were the primary inspiration for such great men as Albertus Magnus, Thomas Aquinas, and Roger Bacon. These and hundreds of other European scholars depended upon the philosophy of Islamic scientists/physicians to guide their thinking. Aristotle and Hippocrates were secondary. Said the French historian Renan concerning the works of Albertus Magnus, "he owed everything to Avincenna" (i.e. Ibn Sina), and regarding St. Thomas Aquinas, "he owed...everything to Averroes" (i.e. Ibn Rushd).120,121

Yet another genius was Baghdad's al-Kindi, who lived in the 9th century. Al-Kindi was one of the most prominent religious scholars of his time. As was the case with the other Islamic giants, he developed skills in numerous fields, including philosophy, physics, astronomy, chemistry, pharmacology, botany, biology, and medicine. As a chemist and physicist he noted that the universe and that which is within it behaves in a systematic fashion, whether these objects are inanimate, such as stellar masses, or animate such as plants, animals, and humans. Among the most brilliant of his statements was this advice to his medical colleagues: "Take no risks, bearing in mind that there is no substitute for health. To the extent to which a physician likes to be mentioned as the restorer of a patient's health, *he should guard against being cited as its destroyer and the cause of his/her death* (italics mine)."122

Ar-Razi was perhaps the most impressive physician/philosopher of the Islamic Era. One of his greatest contributions to medical philosophy was his rule that in order to achieve the cure of disease the state of both the body and mind must be diagnosed and treated. His book, *at-Tibb ar-Ruhani* (Medicine for the Mind and Soul), is a masterpiece in discussing ways to treat the moral and psychological ills of the human mind and spirit. In it he discusses the role played by the frailties of human thinking in the cause and maintenance of disease.

Ar-Razi anticipated Freud by over 1,000 years by determining that certain sexual dysfunctions were due to psychological derangements.[123] He apparently cured a number of individuals of various diseases by treating them exclusively on the psychic level. Even a cursory reading of his works reveals that ar-Razi's theories in psychology and human behavior are more accurate than Freud's. Will it eventually be established that ar-Razi was the founder of the these sciences? Such a study is worth undertaking. Certainly, if accuracy and human progress are the determining factors, ar-Razi is infinitely superior to Freud and/or his contemporaries.

General and Internal Medicine

The accomplishments of the Muslim physicians in this field were vast. They were indisputably the founders of modern general and internal medicine. The less than sympathetic authors of the recently published book, *Medicine: An Illustrated History*, make it clear that the Muslims described and categorized numerous diseases unknown to the ancients. They were the first to establish the institutions of modern hospitals, medical schools, post-medical education, and medical licensure, this also being indisputable.[124]

Internal medicine was the first love for the majority of the great physicians of the Islamic Era. They were fascinated by the workings of the biology and chemistry of the human machine. Thus, they became the first students of human physiology and biochemistry. Anatomy maintained a lesser status, but it was still studied, especially in Islamic Spain.

The basis of Islamic medicine was a blend of the edicts of the Qur'an, the teachings of Muhammad, and the knowledge of medicine from antiquity. Regarding the latter it was by far the Grecian writings which had the greatest impact. However, the Islamic physicians were quick to assimilate rational approaches to the healing arts from every ancient civilization with which they came into contact. Significant contributions arose from Egyptian, Persian, Byzantine, and Indian medicine. The practices of regions as far as China were analyzed. However, the vast majority of their accomplishments were the result of their own original contributions.

The Muslim physicians were the first to clearly delineate the contagious nature of certain diseases. Ravages of that period, such as smallpox, measles, and the Plague, were, for the first time, explained logically and thoroughly. No one so completely and accurately elucidated these diseases as ar-Razi.

Ar-Razi, who lived in the 9th and 10th centuries, was the greatest clinician of his era. More than simply a doctor, he was an accomplished author, philosopher, pathologist, pharmacologist, botanist, psychiatrist, medical educator, toxicologist, and chemist. However, his most prolific accomplishments were in the field of health and medicine. Ar-Razi wrote and edited volumes on these subjects, performing the work of what would today require teams of men. His medical writings are so advanced that much of what he described could be readily applied today.[125]

Clinically, ar-Razi was considered untouchable. His innovations in chemistry and pharmacology remained unmatched for centuries. With amazing thoroughness ar-Razi discussed in his textbook of medicine the causes of numerous diseases. He wrote the first correct descriptions of small pox as a contagious disease, listing its signs, symptoms, and treatment in modern detail. He was the first to distinguish it in terms of symptoms and signs from measles. His treatise on these diseases were the first ever on infectious conditions.thus, ar-Razi was the original physician to systematically describe the signs and symptoms of viral infections. In fact, his writings on smallpox and measles are so comprehensive that little could be added to them today. His books amount to virtual encyclopedias of medicine.

It was ar-Razi who innovated the use of cold water for the treatment of persistent fevers. Interestingly, with acute fevers the treatment was steam baths. Ar-Razi was the first individual to carefully document the causes and symptoms of various types of fevers, a classification which is found currently in medical textbooks. Furthermore, he originated the concept that fever is merely a symptom of disease rather than a disease in itself.

Ar-Razi could have taught today's physicians much about bedside manner. He warned his students about coarse distant attitudes toward patients. Doctors, he felt, must empathize with the patients and give them hopeful comments in order to help them feel better and maintain a positive attitude.

His educational prowess involved much more than simply philosophical advice. Ar-Razi invented and perfected the *case history*, a standard component in the clinical education of today's medical students.126 Hospital rounds as a means of student education must also be attributed to him. This master was surrounded by an entourage of students as well as physicians. Thus, ar-Razi conceived of and initiated continuing medical education 1,000 years prior to its beginnings in the West. Known today by physicians as CME, this system could gain much from applying ar-Razi's methods of hands-on education and clinical participation.

In the hospital setting ar-Razi was a genius in demonstrating proper medical therapies. His bedside diagnostic prowess was legendary. He was a firm believer in the use of tested herbal medicines known to ameliorate specific ailments. However, he also supported the testing of experimental drugs on both animals and patients to prove or disprove their actions. Thus, he was the first to screen drugs through experiments for potential toxicity. His discoveries included antiseptics, cardiac drugs, purgatives, and emetics, that is drugs which induce vomiting.

Ar-Razi discovered the curative value of healthful foods and good nutrition hundreds of years before his modern contemporaries. This perceptive clinician thoroughly understood the psychology of health improvement. He advised his students to allow patients to eat the foods they loved most, as long as they were healthful ones. However, he also stressed the importance of a balanced diet. He warned patients that frequently changing doctors would be a waste of time, money, and health. In an interesting reversal from the modern way this humble scholar admonished physicians to avoid extravagance and to dress, eat, and live simply.

The fact is ar-Razi made the Greek physicians appear as amateurs and would, if alive today, supersede the majority of the current ones. This prolific scholar produced more than 130 books and treatise in his lifetime. It was the famed historian Will Durant who stated that ar-Razi's writings are so progressive that even today's physicians could benefit from them.

In his book, *Introduction to the History of Medicine*, F. H. Garrison implies that ar-Razi was an equal of Hippocrates. What Garrison fails to mention is that many of Hippocrates' medical theories were erroneous. Ar-Razi surpassed him in this regard. He was a clinician who relied primarily on observation and experimentation, not on guesswork as Hippocrates often did. Aristotle, who relied even less upon science than Hippocrates, was far inferior to ar-Razi in the arenas of medicine and health as well as in medical philosophy. It was ar-Razi's book which was used as a primary text for the education of Western physicians for hundreds of years, not Hippocrates.

Ar-Razi left such a profound impact upon the science of medicine that it is still felt extensively to this day. Anyone who carefully analyzes medical history would be obliged to conclude that ar-Razi's role in creating modern medicine was significantly greater than that of Hippocrates.

Late in life ar-Razi was blinded. An ophthalmologist suggested remedial eye surgery. His comment was most interesting; "I have seen enough of this old world, and I do not cherish the idea of suffering the ordeal of an operation for the hope of seeing more of it." Shortly thereafter, the great, grand ar-Razi died. Perhaps ar-Razi knew that his blindness was irreversible. He worked extensively with chemicals, particularly the highly toxic mercury, as well as lead, which causes cataracts as well as blindness.

If ar-Razi was surpassed by any other it was Ibn Sina. Says M.A. Martin in *The Genius of Arab Civilization*, "In any age, Ibn Sina would have been a giant among giants." Taken literally, Martin is proclaiming that if Ibn Sina were alive today, he would compete with, perhaps supercede, the current masters. He is regarded by numerous historians as one of the greatest physicians of all times, and many esteem him as the greatest physician of the Islamic Era.

Ibn Sina wrote libraries on medicine and pharmacology. His *Canon of Medicine* was the basis of medical studies in much of Europe for over 600 years.[127,128] This million word manuscript summarizes all of Grecian medicine, ancient Syrian, Indian, Persian, and Arabic practices, plus it contains many of his own medical notes and experiments. The Canon focuses on the importance of treating disease with natural methods, including dietary modification, herbs, and nutrients, in preference to the reliance on refined drugs.

Ibn Sina made numerous original observations regarding the diagnosis and treatment of disease. His treatise on diabetes was thorough and accurate, and he was the first to become aware that sugar is spilled into the urine. Before him, the most precise description of diabetes was written in the 9th century by Ibn al-Ishizzar, who noted that the symptoms included frequent urination, large volumes of urine, excessive thirst, and heightened appetite; neither the sweet smell or "taste" of urine were mentioned. Ibn Sina also gave the first accurate description of meningitis, a disease formerly thought to be a mental illness. He was perhaps the original true internist, propounding that medicine was the primary treatment for disease and that surgery was inferior as a cure.

Manipulative medicine was one of Ibn Sina's most unique therapeutics. He was one of the originators of spinal manipulation, that is the forcible setting of joints, to correct structural abnormalities. Greek writings indicate that Hippocrates also utilized this treatment. Recently, scientific studies have proven that spinal manipulation is one of the most effective treatments known for back pain and far surpasses surgery in this regard.

Ibn Sina made pioneering discoveries regarding the role of climate and environmental factors in the genesis of disease. The danger of the spread of disease by water, feces, and soil was emphasized. Thus, he was aware of the communicable nature of many diseases. Perhaps most valuable in this regard was his discovery that tuberculosis could be transmitted by infected secretions. These statements make it clear that Ibn Sina was the originator of two modern medical specialties: preventive medicine and public health. His discoveries were instituted at a time when the Western world regarded bathing as blasphemous.

One of Ibn Sina's most astounding observations was that cancer must be diagnosed as early as possible so it can be treated in the initial stages, and, if surgery is to be used, the physician must be sure to remove all of the diseased tissue. This is precisely the modern technique for cancer surgery. He was also quick to note the important role played by the mind in the cause of cancer and other diseases.

As a scholar Ibn Sina was one of the world's greatest masters. A famous physician by the age of eighteen, he was the most prolific medical writer of his time, in fact, of many centuries to follow. By his mid twenties he had already published several books. His greatest feat was his literary output, which was so massive that it completely overshadowed the writings of any other author, ancient or modern. What was most amazing is that all of this was achieved with nothing more than the gift of manual dexterity, hard work, paper, and the fountain pen. Obviously, there were no typewriters, printing presses, or computers in that era. Yet, in what is one of the most monumental achievements in human history Ibn Sina wrote some 270 books and treatise over a 40 year period.129

Ibn Sina's advancements had an enormous impact on the scope and direction of Western medicine. Sarton, the Harvard historian of science, states that Ibn Sina's triumph was so complete that it discouraged others all over the globe, as they could discover little more, and original investigation waned for centuries after his time. His textbook was the basic medical reference in Europe for a longer period than any other book ever written.130,131 Today, a major textbook might remain authoritative for 50 years maximum. In what is an utterly astounding feat Ibn Sina's remained the authority for fully 700 years.132

Singlehandedly, Ibn Sina's writings revolutionized medical practice in the West, replacing sorcery, superstition, folklore, and barbarism with deliberate science. However, barbarism resisted stubbornly, and it was not until the mid-20th century that the practice of medicine in Europe or America became as refined as it was during the era of the Islamic Empire.133-135

There is little doubt that Ibn Sina was superior to Greek physicians. In fact, the impact of his writings, that is his medicine and his philosophy, upon the progression of Western medicine was greater than that of Galen, Hippocrates, and the remaining Greco-Roman physicians combined. This is confirmed by the fact that Ibn Sina's textbook was the primary source and, in some instances, the sole source of medical education in Europe for hundreds of years.[3]

Muslim physicians effectively treated diseases for which the West knew no remedy. Through the prescription of specific herbal remedies and drugs, they cured many individuals with Plague, smallpox, influenza, pneumonia, parasitic illnesses, tuberculosis, heart failure, kidney infections, and measles, diseases which were frequently fatal at that time. They prescribed effective remedies for dysentery and intestinal worms, including purgatives and antiseptics. Sulfur was utilized for the treatment of scabies, an effective and safe preparation.

The Muslims were the first to identify nutritional deficiency as a cause of disease. In this respect goiter, a plague in many parts of Europe, was treated with sea weed as a source of iodine. It was az-Zahrawi who first recognized anemia as a nutritional deficiency. He cured it with inorganic iron.136

Muslim doctors accurately described the medicinal uses for thousands of drugs. Aloe, ginger, and oregano, now highly popular, are a few examples. The word aloe itself is derived from an Arabic word. Additionally, they recognized that the major healing components of aloe were located within the leaf, not the inner gel, a discovery modern chemists have only recently made.137

Muslim physicians used natural medicines aggressively and

---

[3] Ancient records recovered from Louvian University reveal that Ibn Sina's, as well as ar-Razi's, texts were regarded as *the only authentic references until the 1600s.* Their use was continued on a lesser status for many additional decades.

based such usage on sound science. They were the first to test herbal drugs on animals to evaluate physiological effects. This was an epic development, since it marked the beginning of modern pharmacology. It also proves they were exceptionally cautious in administering drugs that might cause serious harm. In this regard they clearly surpass today's doctors, both Western and Eastern. Thus, Islamic medicine was the original scientifically applied system of herbal medicine.

Islamic civilization cultivated the most profound interest in natural pharmacy ever known, far exceeding the accomplishments of the Greeks. Their expertise in the manufacture of various natural drugs has never been surpassed. Muslim chemists, physicians, and pharmacists wrote entire libraries on the subject of natural pharmacology. They knew exactly how best to administer the drugs. They also had a thorough knowledge of the side effects if any. More importantly, they described in detail which diseases they cured. Yet, it is important to note that these scientists also manipulated natural compounds to enhance biological activity. They perfected drug refinement for the achievement of optimal pharmacological effects eight centuries before the West's use of it. Their books are replete with specific instructions regarding chemical methods for drug preparation and/or synthesis. Thus, they must be regarded as the creators of the first pharmaceutical houses. The word drug itself is of Arabic origin.

The Islamic philosophy of health care permeated an entire Empire and, in fact, extended far beyond it. The Muslims' knowledge of medical ethics, diagnostic approaches, patient care, and natural cures reached the most distant lands of the then-known world: China, Korea, Japan, Northern Russia, Scandinavia, South Africa, India, and Indonesia, having a lasting impact upon each of them, an influence which is felt currently. Western medicine itself is a derivative of Islamic medicine. Unfortunately, the former has deviated from the altruistic principles which were carefully maintained by Islam and which were the legacy of its civilization.

References

(see Bibliography for more detailed information)

1. Goldstein, Thomas. *Dawn of Modern Science.*

2. Durant, Will. *Age of Faith.*

3. Sarton, George. *Introduction to the History of Science.*

4. *Dictionary of Scientific Biography.*

5. Turner, H.R. *Science in Medieval Islam.*

6. The Definitive Sayings of Prophet Muhammad (*Sahih al-Bukhari*).

7. Ibid.

8. Bammate, H. *Muslim Contribution to Civilization.*

9. Garrison, F. H. *An Introduction to the History of Medicine.*

10. Lyons, A. S. and R. J. Petrucelli. *Medicine: An Illustrated History.*

11. Goldstein, Thomas.

12. Singer, Charles. *A Short History of Scientific Ideas.*

13. Bammate, H.

14. Goldstein, Thomas.

15. Garrison, F. H.

16. Durant, Will.

17. Taylor, W. *Arabic Words in English.*

18. Singer, Charles.

19. Garrison, F. H.

20. Durant, Will.

21. Hayes, J. R. (ed). *The Genius of Arabic Civilization.*

22. Lyons and Petrucelli.

23. Hitti, P. K. *Makers of Arab History.*

24. The Definitive Statements of Prophet Muhammad (*Sahih al-Bukhari*).

25. Garrison, F. H.

26. Durant, Will.

27. Bammate, H.

28. Garrison, F. H.

29. Goldstein, Thomas.

30. Hayes, J. R.

31. Lyons and Petrucelli.

32. Goldstein, Thomas.

33. Bammate, H.

34. Durant, Will.

35. Hayes, J. R.

36. Durant, Will.

37. Garrison, F. H.

38. Singer, Charles.

39. Bammate, H.

40. Sarton, G.

41. Durant, W.

42. Durant, W. and Goldstein, T.

43. Garrison, F. H.

44. Durant, W.

45. Singer, Charles.

46. Durant, W.

47. Sarton, G.

48. Bammate, H.

49. Ibid.

50. Durant, W.

51. Burk, J. *The Day the Universe Changed.*

52. Bammate, H.

53. Durant, W.

54. *Dictionary of Scientific Biography.*

55. Syed, Ibrahim. Islamic Medicine: 1000 Years Ahead of Its Time. In: Athens, S. (ed). *Islamic Medicine.*

56. Ibid.

57. Goldstein, Thomas.

58. Hayes, J. R.

59. Goldstein, Thomas.

60. Garrison, F. H.

61. Lyons and Petrucelli.

62. Ibid.

63. Singer, Charles.

64. Hayes, J. R.

65. Bammate, H.

66. Lyons and Petrucelli.

67. Campbell, D. *Arabian Medicine and its Influence upon the Middle Ages.*

68. Bammate, H.

69. Garrison, F. H.

70. Ibid.

71. Garrison, F.H.

72. Goldstein, Thomas

73. Lyons and Petrucelli

74. Nasr, S. H.

75. *Dictionary of Scientific Biography.*

76. Nasr, S. H.

77. Athar, Shahid (ed). *Islamic Medicine.*

78. Garrison, F. H.

79. Lyons and Petrucelli.

80. Ibid.

81. Hayes, J. R.

82. *Dictionary of Scientific Biography.*

83. Garrison, F. H.

84. Ibid.

85. *Dictionary of Scientific Biography.*

86. Hayes, J. R.

87. Lyons and Petrucelli.

88. Ibid.

89. Goldstein, Thomas.

90. Mathe, Jean. *The Civilization of Islam.*

91. Garrison, F. H.

92. Goldstein, Thomas.

93. *Dictionary of Scientific Biography.*

94. Lyons and Petrucelli.

95. Goldstein, Thomas.

96. *Dictionary of Scientific Biography.*

97. Ibid.

98. Asimov, Isaac. *Asimov's Chronology of Science and Discovery.*

99. Ibid.

100. Butterfield, H. *The Origin of Modern Science.*

101. Ronan, C. A.

102. Ibid.

103. Garrison, F. H.

104. Ibid.

105. Hayes, J. R.

106. *Dictionary of Scientific Biography.*

107. Boorstein, Daniel. *The Discoverers.*

108. Garrison, F. H.

109. Lyons and Petrucelli.

110. *Dictionary of Scientific Biography.*

111. Bammate, H.

112. Hayes, J. R.

113. Sarton, G.

114. Athar, Shahid. *Islamic Medicine.*

115. Hayes, J. R.

116. Sarton, G.

117. *Dictionary of Scientific Biography.*

118. Hayes, J. R.

119. Ibid.

120. Bammate, H.

121. Renan, E. *Micellany of History and Travel.*

122. Hayes, J. R.

123. Al-Akkad, A. M. *Arabs Impact on European Civilization.*

124. Sarton, G.

125. Ibid.

126. Ibid.

127. Durant, W.

128. Sarton, G.

129. *Dictionary of Scientific Biography.*

130. Sarton, G.

131. *Dictionary of Scientific Biography.*

132. Sarton, G.

133. Hayes, J. R.

134. Singer, Charles.

135. *Dictionary of Scientific Biography.*

136. Ibid.

137. Hennessee,O. M. and B. R. Cook. 1989. *ALOE: Myth, Magic, and Medicine.* Universal Graphics.

CHAPTER 4

Mathematics

The scientific historian Thomas Goldstein aptly describes the Islamic contribution to modern mathematics as "an absolutely momentous one." For those who are unfamiliar momentous is defined as decisive or "very, very important."

The Islamic scientists held the discipline of mathematics in the highest esteem. They quickly realized that mathematics and its various branches formed the basis of the majority of the sciences. In fact, many of their developments in this field were a consequence of their need to utilize it in solving astronomical problems. The Muslims dominated the science of mathematics for five complete centuries. What's more, they assumed the responsibility of teaching it to the peoples of the world. They perfected mathematics, bringing it to "a high level of sophistication."[1] For, as they knew, what would science be without mathematics?

The wisdom of the ancients greatly influenced Muslim scholars as they studied, developed, and expanded mathematics. Ptolemy's astronomical math, Pythagoras' studies in the theory of trigonometry, Persian astronomical calculations, and the interesting Indian development (200 B.C.) of the basic numerical symbols were all utilized. Using this diverse knowledge they constructed the principles which are the basis of modern mathematics. With an enthusiasm unmatched in history Islamic scholars of the 8th through 10th centuries recovered and translated all of the existing works of the ancients concerning mathematics. They then studied these texts intensively with the objective of upgrading the information. The degree of advancement they achieved is illustrated by DeVaux, who writes that the Muslims "became the founders of the arithmetic of everyday life."2,3 To repeat, they were the *founders*. In the monumental modern study of Islamic science, *The Genius of Arab Civilization*, the mathematicians of the Islamic Era are described as the originators of the disciplines of algebra, arithmetic, and trigonometry. Some historians maintain that their works were the stimulus for the creation of non-Euclidian geometry.4,5 This is the type of geometry which Einstein used to create the Theory of Relativity. The creative genius of Islamic mathematics is further emphasized by De Vaux, who says, "they (the Muslims) were *indisputably the founders of plane and spherical trigonometry* (italics mine)."4

What a travesty it is that this is never taught in the school systems. The names of the Arabic/Islamic fathers of mathematics are rarely if ever mentioned in public schools. In fact, quite the opposite is implied, that is that this science is fully of Grecian and/or European origin. Greek writings themselves dispute this. Plato made it clear that much of Grecian knowledge of mathematics was imported from ancient Egypt, where arithmetic and geometry were essentially a cult of the Priesthood.6 The Babylonians also had a major impact upon Grecian knowledge. Their work in astronomy required that they develop expertise in the mathematical sciences. It was the Babylonians who, among other discoveries, made the first predictions of the eclipses, a mathematical achievement often attributed to Greek astronomers.7,8 Even the premier Grecian mathematical development, geometry itself, never advanced beyond a "theoretical art." As for algebra and practical arithmetic, these remained purely theoretical to the Greeks' own admissions.9

The "Arabic" numerals are perhaps the most impressive achievement of Islamic mathematics. This item should cause

---

[4] Plane trigonometry is the study of triangles on a one dimensional surface, while spherical trigonometry is the study of three dimensional triangles.

every teacher, pupil, and historian to comprehend the tremendous impact of Islamic science on Western culture. However, few individuals are aware of the connection.

The Arabic numerals are defined simply as the numbers in use in the Western world today. These are nothing more than symbols and were originally developed by the scholars of India, not the Arabs. Their proponent, al-Khwarizmi (9th century) of Persia, graciously referred to them in his writings as the "Indian Numerals."

The Indians applied their numerals primarily for commerce and had not advanced them for use in the sciences. It was the Arabic-speaking Muslims who originated their usage systematically in arithmetic and then introduced them to the West. In other words, they created a system for using the numerals in everyday life. Then, they popularized the system globally for use not only in mathematics but also in commerce, navigation, engineering, astronomy, geology, and geography. Since it was an Arabic/Islamic introduction, the name was edified. It was entirely a European innovation to call them "Arabic Numerals," and Pope Syllvester II (10th century), a former student at the Islamic universities, was one of the first to coin this term. However, it was Italy's Leonardo Fibonacci and England's John of Hollywood (13th century) who spread the word about the value of Arabic numerals throughout Europe. This was at a time when all of Europe relied upon the Roman numerals, which impeded the growth of the sciences as well as business and industry. What the Muslims did provide as primarily their invention was the missing link: the modern symbol and method of usage for zero (0). This sign, an essential component of modern mathematics, was established in 976 A.D. by Muhammad bin Ahmad. The fact is zero is a derivation of the Arabic term, *sifr*.[10,11,12]

The linguistics regarding the origins of zero illustrate the tremendous impact of Islam upon modern mathematics. It began with the Arabic word *sifr*, which means "empty." This was translated during the 12th century into Latin as "zephyrum." Thereafter, it was shortened to "zephir," which is found in today's dictionaries as "cipher." Italian mathematicians eventually abbreviated it to its current form.

The introduction of the zero revolutionized math, at least from the Islamic point of view; it was not used in the West until 400 years later. Writes De Vaux, "By using ciphers" (another word for zero and the closer transliteration of sifr) "the Arabs became founders of the arithmetic of everyday life; they made algebra an exact science and developed it considerably...The (Muslims) kept alive higher intellectual life and the study of science in a period when the Christian West was fighting desperately with barbarism."[13]

Scientific historians have determined that the entire number theory in use today is an Islamic advancement. During the 9th through 13th centuries the study of mathematics and arithmetic was required in every school and university within the Islamic Empire. Each mosque, school, and university maintained an extensive library on the mathematical sciences. Additionally, Muslim scholars were the first to apply mathematics to a wide range of sciences. They edited and wrote thousands of books containing mathematical wisdom over a period of some 500 years. This was certainly ample time for elevating the mathematical sciences to a state of perfection. During that same period literary output in Europe in these sciences was nil.[14,15]

Prior to the introduction of the Arabic numerals Europe was in a quagmire and was unable to advance the sciences. The cause of this stagnation was the unrelenting use of the Roman numerals. This was an extremely crude, cumbersome system for counting. Its use inhibited the advancement of the sciences, which are largely dependent upon mathematics. Arithmetic dependent sciences include physics, analytical chemistry, engineering, biochemistry, astronomy, medicine, pharmacology, geography, geology, and geometry. The Greeks used a system of math based upon ancient Middle Eastern arithmetic, which encouraged to some degree advances in the sciences. However, the archaic Roman numerals suffocated any chance for the rebounding of ancient science. These numerals were a yoke upon the necks of Europeans of such an enormity that they completely choked the development of the sciences for some 700 years. Paul Tannery states of geometry in eleventh century Europe: "This isn't a chapter in the history of science—it is the study of ignorance."

To its disadvantage Europe continued to resist the introduction of the Muslim developed numerical and decimal system for several hundred years, stubbornly clinging to the clumsy Roman numerals instead. However, its scholars eventually incorporated them beginning in the 13th century as a result of Latin translations of Arabic works, although the widespread usage of the Arabic numerals did not occur until many centuries later. This had to be so. Without the Arabic numbers and the zero the West's great advances in the sciences would have been impossible. Thus, the systematic application of the Arabic numerals and the zero within European society finally shattered the death grip of Rome. As a result, wisdom, science, technology, philosophy, and civilization began to flourish in a land which was suffocating in an intellectual vacuum. Finally, the European mind began to question the "logic" of the ancients.

Western scholars eventually realized that the great Arabic numerals plus the zero were exactly the tool Europe required to modernize. This was because these numerals were the perfect system for simplifying mathematics, creating, as Islam did, "infinite" possibilities in the fields of commerce, science, architecture, and industry.

There was yet another important breakthrough that was required before civilization could be rapidly modernized: the decimal. It was al-Uqlidusi and Abu'l Hasan who in the 10th century first conceived of it. Yet, it was not until the 15th century that decimals were applied to both whole numbers (for instance, 1.2) and fractions (for instance, 0.2). This was accomplished by al-Kashi in his book, *Key To Arithmetic*. The invention of the decimal was absolutely critical for the transformation of math, arithmetic, algebra, and calculus into true modern sciences.

The invention of practical algebra was perhaps the most pronounced achievement of the Muslim scientists in mathematics. The term itself is a derivative of the Arabic *al-jabr*, which means the reduction of complicated numbers to simpler symbols. That name was attached to it as a result of the title of its inventor's book, al-Khwarizmi's *Al-Jabr wa'l Mukabala*.

Through the efforts of al-Khwarizmi, who worked in the 11th century, algebra was first distinguished from geometry as a separate science. Al-Khwarizmi developed the number concept into our modern one, causing the usages of the Arabic numerals to be infinite. In contrast, it is essentially impossible to reach infinity or anywhere near it with Roman numerals. Yet, the key component of al-Khwarizmi's works was his use of algebraic equations for the simplification of cumbersome math problems. This is best described by al-Khwarizmi himself, who wrote that his objective was simply to show "what is easiest and most useful in arithmetic."[16]

Al-Khwarizmi has been rightfully called the most accomplished mathematician of the Middle Ages. Hitti claims his was the greatest influence on Western mathematics.[17] This is evidenced by the fact that al-Khwarizmi's book served as the basic text for mathematical education in Western Europe for nearly 400 years, a period nearly twice as long as the existence of the United States. That is an incredibly long time for one text to serve as the authority, a feat unmatched in history.

However, in the 12th century Omar Khayyam developed even more complex equations than Khwarizmi's. Incredibly,

Khayyam solved equations of the third and fourth degrees by using intersecting conics.[5] This was the highest algebraic achievement of that era and the highest achievement in modern mathematics when solving equations of the fifth degree or higher.[18] It is important to note that the techniques found in Khayyam's book for solving equations are the same ones used for teaching today's school children, an impressive feat considering that his works were published over 700 years ago.

While virtually all history textbooks attribute the discovery of the binomial theorem to Isaac Newton, it was, in fact, an Islamic discovery. Sir Lancelot Hogben describes in *Mathematics in the Making* that while Western textbooks "loosely speak of Newton as its discoverer," he wasn't. Rather, it was Omar Khayyam who during the 13th century made the first thorough description of it, and Hogben provides an illustration of Khayyam's manuscript to prove it. The binomial theorem is one of the most integral of all mathematical formulae, constituting the very basis of much of modern algebra. In what is yet another monumental error by modern historians the system established to simplify the use of the binomial theorem is known alternatively as the Chinese triangle or Pascal's triangle, the latter referring to its supposed founder, Blaise Pascal of 17th century Europe. Yet, neither the Chinese nor Pascal discovered it. It was again the brilliant Khayyam who first described and published the system, 500 years before Pascal.[19] Therefore, it must be renamed "Khayyam's Triangle."

During the 15th century Islamic Persia's al-Kashi made a monumental contribution, advancing algebra far beyond what was previously known. According to the *Dictionary of Scientific Biography* he wrote a "high quality" practical encyclopedia of mathematics that had a "clarity and elegance" which matches,

---

[5] A conic is a one-dimensional section of a cone, i.e. an ellipse, hyperbola, parabola, etc.

perhaps exceeds, the finest modern texts. Within it he gave comprehensive methods for solving equations, which are precisely the methods in use today. The text was used as an authority for hundreds of years and had a major influence upon European mathematics.

Al-Kashi was perhaps the ancient world's greatest mathematical genius if not the greatest of all time. He wrote an entire book regarding the precise calculation of the circumference using higher mathematics. His work is such an elegant example of advanced mathematics that the *Dictionary of Scientific Biography* describes it as "a masterpiece of computational technique."

It is difficult to quantify the immense achievements of the Muslims in the field of algebra. The equation, which is the basis of this science, is their innovation. They perfected the binomial theorem and introduced it to the West. Much of algebraic expression is based upon this theorem, and it is taught currently in elementary and secondary schools all over the world. They were the first to use algebraic symbols, so commonly applied today for teaching this science. The concept of the positive (+) and negative (-) signs was introduced to the West by Islamic mathematicians. The modern term for multiplication, "x," is likely a derivative of a figure used by Islamic mathematicians.20 The Muslims also invented logarithmic tables. Additionally, irrational numbers are an Islamic innovation and were introduced by Kamal ad-Din in the 13th century. The square root of 2 is an example.21

Virtually all types of algebraic equations are Islamic inventions, since the Muslims were the first to use symbols to represent the "unknown" variables. It is probable that the square root symbol in use today is a "Latinization" of the Arabic letter *jim*, which was the one used by al-Khwarizmi as well as other Muslim algebraic experts.

All mathematical sciences are dependent upon algebra: for

instance, trigonometry and geometry. Thabit bin Qurrah was the first scientist to determine that algebraic equations could be applied to geometry in order to modernize the science.[22,23] The impact of algebra upon the direction of science is also emphasized by the discoveries of al-Battani, who modernized and perfected trigonometry through the use of algebraic equations.

The current terminology in algebra is dominating evidence of the influence of Islamic science. The algebraic term algorithm, and, therefore, logarithm, which is defined as "a system used for solving of mathematical problems" is Arabic-derived. The source is the Latin version of the last name of the greatest mathematician of the Middle Ages: "Algorismus," i.e. al-Khwarizmi. What's more, a similar term, *algorism*, which is also derived from his Arabic name, was in fact the Latin word for arithmetic throughout Europe for nearly 400 years.

The Islamic scientists were diligent specialists in the field of trigonometry, which is defined as the study of triangles. This was largely a consequence of the intimate relationship between it and astronomy. Additionally, they became highly proficient in geometry. Their esteemed accomplishments include the solving of cubic equations (i.e. x to the third power).[24,25] Cubic and rectangular equations form much of the basic learning of today's grade schoolers in this science. Additionally, Muslim scholars performed extensive studies on cones and cylinders, laying the foundation for the modern analysis of these figures. The methods children use today to analyze them are Islamic derived. Many of the advancements in geometry and trigonometry originated from the works of al-Battani, one of the most prominent students of mathematics of his age and the first individual to propel trigonometry into the realm of independent science.[26]

During the 11th century al-Battani conceived of methods to improve the measurements of angles used by the Greeks. The latter utilized chords to measure the arcs of triangles. This method was cumbersome. Al-Battani introduced instead the sine of the arc. Sine was used in ancient India, as was cosine. However, according to M. Charles and others al-Battani was the first to use these expressions in developing the science of trigonometry.[27,28] C. Ronan says al-Battani also innovated the application of trigonometry for the systematic study of other sciences, notably physics and astronomy. Yet, in the 11th century Persia's al-Biruni independently accomplished a similar feat by using precision trigonometry in the study of geology, geodesy, astronomy, and physics.[29,30]

Muslim mathematicians are the sole inventors of the remaining basic trigonometric function, the tangent, and coined this term. Later, they innovated the cotangent. This was the climax in the development of trigonometry. It was some five centuries before the West adopted these final pieces of the trigonometric puzzle. As is confirmed by the editors of the scholarly work, *The Genius of Arab Civilization*, the Muslims were the developers of the majority of trigonometric functions and ratios in use today. These functions, the sine, cosine, tangent, and cotangent, are the very basis of modern trigonometry. These facts are compelling proof that the trigonometry in use today is entirely an Islamic innovation.[31,32]

Geometry was also renovated by the Muslims. During the 8th-9th centuries Thabit bin Qurrah innovated the concept that algebra could be used to describe the ratios between geometric quantities. His work elevated geometry from mere theory to a science with practical applications. Surveying, construction, architecture, and artistic endeavor were revolutionized as a result of these outstanding discoveries.[33,34]

The renowned astronomer Abu'l Wafa' was perhaps the greatest geometer of the Islamic age. In addition to his excellent manual on basic arithmetic he wrote an influential treatise on simple/practical geometry. His work was so practical that C.

Ronan claims it "would have upset the Greeks." Abu'l Wafa's book describes the use of the math compass (he invented the one with the moveable arm) and the ruler for solving two and three dimensional geometric problems. Thus, he simplified the complicated geometry of the ancients so that virtually anyone could apply the science. In fact, the geometry in use today is more closely the geometry of Abu'l Wafa', the inventor of the adjustable compass and common ruler, than it is the geometry of the ethereal Euclid.

The current system of mathematics is essentially the same as the one which was laboriously constructed by the Middle Age Muslim scholars. They simplified and perfected algebra. There is very little difference between what is used today and what they used ten centuries ago.35-37 They invented practical geometry.38 They created trigonometry, a science barely developed in Greece.39,40 They produced original findings in the field of integral calculus.41 Their unwillingness to accept Euclid at face value caused them to initiate the development of non-Euclidean geometry.42

The inescapable conclusion is that it was the Muslims who were the founding fathers of mathematics, not the Greeks. De Vaux, Sedillot, Humboldt, Briffault, Renan, and Sarton all emphasize this fact. Numerous basic principles of arithmetic, algebra, trigonometry, and geometry are solely their innovations. The Greeks developed the first accurate principles of geometry, and in trigonometry they offered the Pythagorean theorem. However, in the remaining mathematical disciplines they stopped at elementary theories. Plato himself documented the indifference of his people towards the study of mathematics, noting that the Phoenicians and Egyptians were much more inclined towards it. In fact, he reveals evidence that much of Grecian math was copied from the works of Egyptian priests.43 In an era devoid of calculators or computers Islam's scholars painstakingly perfected the mathematical sciences into their modern form. This was perhaps the greatest achievement of Islamic science.

Doubting Thomas'?

It is realized that the contents of this book is the opposite of what is currently taught in history and math classes. For reasons yet to be determined today's math teachers entirely omit the Islamic contribution. It is astounding that such a blatant oversight could be made. Thus, it is understandable that certain individuals might be shocked by this data. Yet, what is perhaps more shocking is the fact that such an impressive array of data regarding the origins of modern mathematics and other sciences has been withheld, despite the fact that it is so thoroughly documented. It is reprehensible that billions of individuals have been specifically taught false history, largely to disguise the contributions of Islamic scholars. A vast effort has been made to alter public knowledge, but what is the motive? Perhaps it is the fear that by eliminating the most foul and detrimental human behaviors Islam will rise again to infuse this world with glory and achievement. The fearmongerers wish to control humanity by manipulating its destiny, strictly to achieve their own selfish desires. Perhaps the fabrications are motivated by greed, that is a desire to maintain power and control. Whatever the cause, historians are obligated to correct the public record, so that the true evolution of history is available to the world.

It is expected that certain individuals will doubt the findings of this book even though the information is presented factually and is supported by the appropriate references. Even individuals of Arabic and/or "Muslim" extraction might encounter a degree of skepticism or, perhaps, might be overwhelmed when confronted with this data. To a degree doubt is an expected consequence of previous experience, especially if the individual is unfamiliar with the subject matter.

Virtually everyone in the Western world is being continuously bombarded with misinformation—even lies—regarding the history of science. As a result, doubt and/or skepticism would almost be natural. What's more, individuals frequently resist new ideas or facts which are in contradiction to established beliefs. Yet, there are people of the opposite mentality, that is those who are constantly seeking the truth and who spend their lives searching for it. For them learning new or unexpected information is an adventure.

Those who have been taught a false model of history can alter their beliefs. Yet, this can only occur if the individual approaches the issue in question with an open mind. Skepticism and cynicism completely impede any opportunity for free thought. As a result, the individual will likely maintain a biased view of life in general as well as the history of science. Thus, Galileo will always be the father of astronomy, Nicolaus Copernicus, the founder of the correct planetary model, Roger Bacon/Francis Bacon, the father of the "scientific method," Leonardo da Vinci, the first to conduct scientific experiments, Johannes Kepler, the first to establish astronomy as an exact science, Antoine Lavoisier, the founder of chemistry, Isaac Newton, the sole originator of physics, Osler, Galen, and Hippocrates, the fathers of modern medicine, Robert Boyle, the first scientific chemist, William Harvey, the discoverer of the systemic circulation, Galen, the father of anatomy, Joseph Lister, the discoverer of antiseptic surgery, Crawford Long, the originator of anesthesia, Paul Ehrlich, the inventor of drug therapy, John Dalton, the announcer of the atomic theory, Charles Lyell, the founder of modern geology, Diophantus, the father of algebra, Pythagoras, the originator of trigonometry, etc., etc., etc.

References

(see Bibliography for more detailed information)

1. Stewart, Desmon. *Early Islam.*

2. DeVaux, Carra. *The Philosophers of Islam.*

3. Ronan, C. A. *Science: Its History and Development Among the World's Cultures.*

4. Bammate, H. *Muslim Contribution to Civilization.*

5. *Dictionary of Scientific Biography.*

6. Al-Akkad, A. M. *The Arab's Impact on European Civilization.*

7. Ibid.

8. Durant, W. *Age of Faith.*

9. Hayes, J. R. (ed). *The Genius of Arab Civilization.*

10. Ibid.

11. *Webster's New World Dictionary.*

12. Taylor, Walt. *Arabic Words in English.*

13. DeVaux, Carra.

14. Hayes, J. R.

15. Burk, J. *The Day the Universe Changed.*

16. Ronan, C. A.

17. Hitti, P. K. *History of the Arabs.*

18. Hayes, J. R.

19. Hogben, L. *Mathematics in the Making.*

20. Ronan, C. A.

21. Hayes, J. R.

22. Bammate, H.

23. Hayes, J. R.

24. Ibid.

25. Hoyt, E. P. *Arab Science.*

26. Ronan, C. A.

27. Al-Akkad, A. M.

28. Bammate, H.

29. Nasr, S. H. *Science and Civilization in Islam.*

30. *Dictionary of Scientific Biography.*

31. Sarton, George. *Introduction to the History of Science.*

32. *Dictionary of Scientific Biography.*

33. Hayes, J. R.

34. Ronan, C. A.

35. Bammate, H.

36. Ronan, C. A.

37. *Dictionary of Scientific Biography.*

38. Ronan, C. A.

39. Butterfield, H. *The Origin of Modern Science.*

40. *Dictionary of Scientific Biography.*

41. Ibid.

42. Ibid.

43. Al-Akkad, A. M.

CHAPTER 5

Geology and Geography

Geology and geography are sciences which deal with the study of the earth, that is it's surface structures, contents, formations, history, inhabitants, and origins. Geology is defined more specifically as the study of the origin of the earth's inorganic matter, its rocks, mountains, volcanoes, crust, inner core, canyons, valleys, oceans, seas, fossils, and craters. Geography is the study of the surface of the earth, the atmosphere which encases it, and the organisms which inhabit it. Without living creatures, there wouldn't be a science of geography.

Geology

As with astronomy the Muslim scientists had enormous motivation for studying this science. The Qur'an delves into the realm of the structure of the earth. Geological transformations and mechanisms are frequently mentioned: for instance, the interior of the earth is described as in a "state of turbulence." The topography, that is the structure of the earth's surface, is emphasized, for example, "We spread out the earth's crust for you as (a carpet)."

The Prophet Muhammad continuously provoked the thinking of his followers in regards to the earth sciences, primarily with his edicts on seeking knowledge: "Study from birth to death, continuous learning is a duty, seek knowledge even as far as China," etc. While the Prophet's comments might appear symbolic, they are, in fact, serious and deliberate. China was an incredibly long distance to travel via camel, horse, donkey, or pair of legs.

Yet, his statement emphasizes how important he felt the accumulation of knowledge is; so important that if traveling to China would further its cause, one should do so. It should be reiterated that this was at a time when there were no automobiles, trains, airplanes, or other forms of rapid transit. However, Muslim scholars believed they should spare no hardship to accumulate knowledge from the world about them. By other edicts, such as "scholars are the prophets of the future" and "give me a group of men discussing knowledge and another engrossed in worship and I'll join the discussion group" (which he did), Muhammad, in his immense wisdom, placed knowledge itself as the crux of faith.[1] As a result of this philosophy Islamic civilization was highly progressive, converting all of the branches of ancient knowledge, including geology and geography, into scientific endeavors. This emphasis upon knowledge was precisely the opposite of the attitude in Medieval Europe, where the concept of faith alone ruled. While the faith proponents were hopelessly mired in intellectual stagnation, the knowledge seekers of Islam built an utterly impressive civilization, permanently altering the course of history. Says the French historian Renan, "*no people*...contributed so much as (the Muslims) to broadening man's conception of the universe and to giving him an *exact idea* of the planet on which he lives, which is the *prerequisite of all real progress* (italics mine)." Thus, the Muslims were the first individuals in history to turn the study of geology and geography into a science.

Muslim scientists were fully aware that the earth is millions of years old. Furthermore, they knew that throughout these eons the topography of the earth had gradually changed. For the first time in history they described in detail how these changes were induced by geological phenomena such as wind, water, earthquakes, and volcanoes.[2,3] While hundreds of Islamic scholars contributed to this effort, the works of Ibn Sina became the most well known in Europe. Ibn Sina proposed modern theories for the origin of the earth. His works were the primary source for the European advancement of this science. For instance, he was the first to construct a correct theory for the formation of sedimentary rock.4 For the first time in history Ibn Sina accurately described how the earth's crust was formed as the result of gradual actions of natural forces, building layer upon innumerable layer, over a period of eons, a concept which apparently directly influenced Leonardo da Vinci some 500 years later.

Al-Biruni also made a number of original contributions to geology. He produced scientific explanations for the origins of rock formations, precious stones, and minerals. He clearly knew, hundreds of years before it was discovered in the West, that massive geological events, such as intense heat and pressure, were responsible for the formation of precious stones and metals.

Al-Biruni was also one of the world's first makers of quality geological maps. Using complex mathematics he made precise measurements of the earth's topography. For instance, he accurately measured the heights of mountains through mathematical formulae. He also correctly assessed the geological causes for their origin.5

Al-Biruni was a student of fossils. He made the astounding 10th century statement that fossils found on mountain crests indicate a watery origin of the earth. As an example of his deductive reasoning his observations regarding fossils and rock formations led him to conclude that the Indus Valley was once the bottom of a primordial sea.6 Thus, he was aware that the earth had existed for many eons, a relatively common belief among Islamic geologists of that era.

The Arabic word used in the Qur'an to describe the earth's creation, *ayyam*, can be interpreted as an infinitely long period of time (i.e. an eon).7 In contrast, Christian dogma maintained that God created the earth literally in six days and that she is no more than 6,000 years old. Even today people believe in this theory, one which has no scientific basis and can be easily refuted.

A number of Muslim geologists collected fossils and not just as a hobby. They studied and classified them in a manner that would amaze modern geologists.8

Numerous Muslim geologists propounded the theory of the roundness of the earth. Many noted that it was more egg shaped than round. Durant claims that Muslim scholars simply took the earth's sphericity for granted. As an example of this thinking during the 9th century Ibn Rasta commented that God "made the sphere like a round ball." He went further to indicate that it was solid rather than hollow, the latter being a common belief of the ancients.9

Ibn Sina must be regarded as the one of the founding fathers of geology. His amazingly accurate description of the geological origin of the mountains was developed at a time when Europe was encumbered with creationism. Possibly it was Ibn Sina's familiarity with the Qur'an—he knew the book by heart—which lead him to arrive at his theory. The Arabic word in the Qur'an for describing the base of a mountain is defined as a "stake (or peg) driven into the ground," indicating the existence of the deep subterraneous geological folds.10 To quote Ibn Sina:

> "Mountains may be due to two different causes. Either they result from upheavals of the earth's crust, such as might occur in a violent earthquake; or they are the effect of water, which, cutting for itself a new route, has denuded the valleys. The strata are of different kinds, some soft, some hard; the winds and waters disintegrate the first kind, but leave the other intact. It would require a long period of time for all such changes to be accomplished...but that water has been the main cause of these effects is proved by the existence of fossil remains of aquatic animals on many mountains." 11

What is most impressive about these observations is the fact that they were made in the 11th century A.D., some 700 years prior to the supposed birth of modern geology. More importantly, his words prove that he used purely scientific thinking in the analysis of geological formations. Ibn Sina made another tremendous advance when he propounded the most accurate ancient idea for the formation of precious metals. He proposed that they were formed as a result of becoming liquefied from periods of intense heat followed by cooling. This is exactly the modern thinking. As a result of these novel findings regarding the gradual nature of mineral formation Ibn Sina, along with al-Biruni, must be regarded as the founder of the science of mineralogy. In closing, it was the orientalist F. H. Garrison who, in his book, *Introduction to the History of Medicine*, admitted that "the... (Muslims) were themselves the originators...of.. geology."

Geography

The accomplishments of the Muslims in the field of geography were vast. Indeed, they originated this discipline as a science for the study of the globe, resurrecting it from it's Medieval grave of ignorance.

Other than the works of Ptolemy there were no scientific geographical treatise in existence within the Western world during the Middle Ages. Ptolemy's text itself only became available as a result of the Arabic version's translation into Latin complete with its Islamic commentary. Yet, even this was largely unavailable until the 16th century. As a result, Medieval Europe knew nothing of the world about it. Fables and fairy tales abounded everywhere.

During the Middle Ages the geographers of Islam wrote hundreds of thousands of pages about the lands and peoples of the world. They published the world's first almanacs, atlases, and encyclopedias.[12-14] During the 9th century Ibn Khirdazbah constructed the world's original road map atlas. Later, Muslim geographers created relief maps, complete with color coding.[15] These pioneering works served as the prototypes for the maps produced today.

Unfortunately, the vast majority of Islamic encyclopedias have been lost or were destroyed during invasions. However, one that has survived is Yaqut's (d. 1229) *Mu'jam al-Buldan* (Dictionary of Countries). It is a vast geographical encyclopedia, which provides an overview of nearly all of Medieval knowledge about the earth. Interestingly, Yaqut was Greek and was formerly a slave as a consequence of being captured in warfare with the Muslims. However, under the tutelage of Islam he was given an excellent education and was later freed. As yet another example of the model civilization of the Islamic Empire, incredibly, Durant writes that Yaqut the Greek *and former slave* rivals al-Idrisi as "the greatest medieval geographer."

It must be remembered that during this era the whole of Europe understood the world to be flat, a belief supported and maintained by the Church. Those who attempted to prove otherwise were persecuted, imprisoned, tortured, or murdered.16-18

The scholars of Europe first learned of the earth's roundness from Latin translations of Arabic works. Thus, the Muslim geographers provided the major impetus for altering Europeans' views regarding the world about them. They, not the Greeks, are fully responsible for disseminating into the West the roundness theory for the earth as well as for the sun, the moon, and the remaining planets.19-23

The Explorers of Islam

There were no airplanes, blimps, helicopters, or space ships during the Middle Ages, at least none that we know of. Geography could only be studied and mapped by traveling about the globe by foot, horseback, or ship. Says Renan about the Muslims, "their passion for travel is one of the most striking traits....and one of those which have helped them to make their *deepest mark on the history of civilization* (italics mine)." Furthermore, Renan indicates that Muslim scholars broadened man's perspective of the universe to a much greater extent than did the Greeks or any other ancient society. Yet, little or no mention of this significant accomplishment is given in standard textbooks or encyclopedias on the subject. *Encyclopedia Britannica* does some justice regarding the Islamic contribution. The editors note that Muslim geographers journeyed far more extensively than any previously. For instance, they describe how Ibn Battutah during the 1300s landed as far south in Africa as what is now known as Ghana. The editors note that Ibn Battutah's explanation of the moderate equatorial climate was the first to displace the erroneous presumptions of Aristotle, who thought that the region was not inhabitable. "Ibn Battutah," they write, "covered 75,000 miles, ranging as far east as India and the Malay archipelago (i.e. the Malaysian peninsula)." That is an incredibly long distance by any travel standards. It is comparable to thirteen trips back and forth across the United States. Yet, the Britannica calls this brilliant geographical historian, a man who was perhaps the most learned explorer of the Middle Ages, a "Muslim merchant." This while Sir Percy Sykes makes it clear that the scholarly Ibn Battutah was a geographer in the purest sense, calling him "the greatest of all Muslim explorers."24

It seems that English books have difficulty making positive comments about Islamic accomplishments without making correspondingly derogatory ones. Thus, while these previously mentioned comments on Islamic geography are somewhat complimentary, the editors of the Britannica open and close the geography section with negative statements and, perhaps more importantly, negative innuendos. "Muslim Geographers—The dark age of geography began...." is their opening. They close with, "The Muslims, however, contributed nothing to the progress of cartography" (i.e. mapmaking). Durant's writings alone thoroughly dispute this. He states that al-Idrisi's maps "were the crowning achievement of Medieval cartography, *unprecedented in fullness, accuracy, and scope* (italics mine)."

While many people believe that the words of the Britannica are gospel, this book has proven the editors wrong on numerous counts. Is not the precise determination of latitude and longitude a crucial component necessary for the advancement of map-making? The Muslim scholars corrected extensive errors by the Greeks in this regard and in some instances measured latitudes and longitudes nearly as accurately as they are known today.[25] What about Islam's introduction of the first globes?[26,27] Is the production of untold thousands of books and monographs on the historical geography of the lands and peoples of the world, descriptions which Dunlop and Sykes describe as "unmatched," to be equated with a "dark age" in this science? The fact is such an output of original data in geography remained unequaled until the 20th century. The editors of the Britannica make a monumental error by equating the enlightened era of Islamic geography with the dark ages. It took Europe nearly 600 hundred years from the time of Ibn Battutah to equal the literary accomplishments of Islam's scholars during what must be regarded as the "Bright Age" in geography. In fact, many of the Islamic descriptions have yet to be equaled.

The misinformation continues under the section entitled "Revival" of geography, where the Britannica's editors claim, "Roger Bacon, writing in the 13th century, is generally credited with originating the modern scientific method." They place him in the Revival section, while Ibn Battutah, who during the 14th century traveled the world over and whose chronicles had a major influence upon European scholars, is placed in the "dark ages." Furthermore, this ubiquitous claim concerning Roger Bacon has been proven false several times within this book. Indeed, it was Durant, among others, who clearly stated that Bacon owed virtually everything he knew—even his philosophy—to Muslim scientists.[28-30]

The Britannica continues its blunders by proclaiming that the Revival was the result of European authors, who "went back to Ptolemy," while admitting in the dark age section that Aristotle and, for that matter, Ptolemy, disseminated misleading information that repressed the development of correct geographical sciences for hundreds of years. "Petrified" was the word used by Durant to describe the negative impact of Ptolemy's geographical and astronomical blunders upon the progress of European thought. Thus, the editors failed to recognized the enormous impact of Islamic geography upon the Western world. Muslim geographers propelled geography into becoming an exact science by their brave criticism of the "authority of the ancients." This was the beginning of the Revival.

Muslim geographers were the first to explore distant regions of the earth. They visited southern Africa and the far north of Russia, even exploring Siberia. They entered China 400 years before Marco Polo.[31] Indonesia, Malaysia, Korea, the Philippines, Ceylon, Central Africa, and Madagascar were visited and explored. They made a surprise visit to Ireland and ventured as far west in the Atlantic as the Azores and, if al-Mas'udi's account is to be believed, perhaps further.[32]

The weight of the credit for the renovation of geography must be given to the Moroccan/Sicilian scholar al-Idrisi. This 12th century genius, who was a direct descendent of the Prophet Muhammad, was one of the first men to apply scientific methods in the study of geography. Prior to him geography in Eurasia was little more than a collection of fables and outrageous stories.

Throughout the Middle Ages no accurate histories of the earth and/or its peoples were produced in Europe, with the exception of Marco Polo's, nor were any reliable maps readily available. The Islamic geographers altered the course of history by changing the way people think. They converted global ignorance and superstitions regarding the peoples of the world into realism.

In Europe, where the populous was completely ignorant about outside cultures, the Islamic geographers and travelers had a ready audience. They enlightened the Europeans regarding the world cultures, drastically altering their jaded views. Europeans, encumbered by illiteracy and intolerance, had no interest in exploring other civilizations, that is not until their encounter with Islam. All European exploration arose after the assumption of the Islamic civilization in Spain during the 15th century. The fact is the scholars of Islam were directly responsible for stimulating the Age of Exploration.[33] This is evidenced by the fact that virtually all of the nautical and geographical terminology used during the Middle Ages was in Arabic. The instruments were of Arabic design. The maps were Arabic. So were the astronomy texts and star charts as was the knowledge of the sphericity of the earth. The longitudes and latitudes, so crucial for global navigation, were calculated in Arabic. What's more, the geographers, whose maps were admired and copied by Europeans for hundreds of years, were Arabic-speaking. Al-Idrisi's influence upon Europe was particularly direct: he was King Roger II of Sicily's lead scientist.

Al-Idrisi's map was the one which inspired Columbus to risk a voyage to the New World.[34,35] It was also al-Idrisi who determined the sources of the Nile river and the cause of its floods. Its sources, he noted, were the lakes located south of the equator, which he illustrated on his maps. Since ancient Egyptian times geographers, including the Greeks, had been at a loss in determining the source of the Nile's headwaters in an effort to comprehend methods of flood control. They wrote that the water came from springs in the southern Nile, which, of course, is erroneous. Yet, the editors of the Encyclopedia Britannica have the gaul to write that the "Revival" was in going back to the archaic Ptolemy.

This great Islamic thinker went beyond Ptolemy by being the first geographer to apply scientific methods for the study of the topography of the earth. Therefore, he may be regarded as the founder of geography as we know it today.

Al-Idrisi was the Middle Age's top cartographer. He drew accurate maps of all of Europe, parts of Africa, and much of Asia. He was admired and respected throughout the Islamic Empire. Students from distant lands came to study with him. This admiration melded into Christendom. In the 12th century Sicily's King Roger II sent for al-Idrisi and attached him to his court. There, in 1154 A.D. he completed the construction of a mapped globe as well as a disc shaped map of the then-known world. Both were made of solid silver at the King's request, and al-Idrisi was paid handsomely with more of the same. Said Sedillot of al-Idrisi's maps, "For 350 years, European cartographers did nothing but copy...(them)..with negligible variations."36 Thomas Welty writes in his textbook of world history that al-Idrisi's maps showed the world with "great accuracy and detail."37 Will Durant notes that the scholarly al-Idrisi wrote an immense multi-volume geographical encyclopedia, "the greatest of its time," which fully described the lands and peoples of the Middle Age world from Spain, Europe, and Africa to the far reaches of Asia. Vasco da Gama, the first European to round the of the horn of Africa, used a derivation of al-Idrisi's map as a guide and procured a Spanish Muslim as his navigator.38 However, even more profoundly, if Sedillot's claim is taken literally, al-Idrisi's map must have been the one used by Christopher Columbus to discover the New World. This is delineated by the modern British historian Dunlop, who writes that Muslims may "claim a share in the discovery of the New World."39 Yet, Columbus utilized far more from Islam than merely al-Idrisi's map. Islamic science was also the source for his celestial globe, star charts, astrolabe, quadrant, sextant, compass, cloth sails, rudder, and magnifying lenses, all of which he relied upon to make his discovery. It was the source for his almanac (Ar-*al-manakh*), his alidade for his sextant (Ar-*al-'idadah*), and his caliper (Ar-*qalib*). It gave him the names of the stars necessary to guide his course in the difficult night skies: Aldebaran (Ar-*aldabaran*), Betelguese (Ar-*bayt al-jauza*), Mizar (Ar-*mi'zar*), Altair (Ar-*al-ta'ir*), Rigel (Ar-*rijl jauzah al-yusra*), Vega (Ar-*waqi'*), and Tauri (*karn al-thaur*) among others. It provided him the mathematics required to complete numerous successful journeys: his algebra (Ar-*al-jabr*) and his trigonometric functions, the sine and cosine (Ar-*jayb*), so he could calculate the zenith or summit (Ar-*as-sumut*) and nadir (Ar-*nathir as-sam't*) to determine his precise location.

America was discovered within months (1492) after the fall of Islamic Spain (1492), a fact which clearly illustrates the dominant role of Islamic science. Islamic Spain was Europe's sole repository of learning, the source of its only public libraries, and the exclusive source of its credible universities. In other words, Islamic Spain, as well as North Africa, was where the scholars of Europe learned the exact sciences: astronomy, geography, mathematics, meteorology, and navigation. This too was where, whether directly or indirectly, Columbus, Diaz, and da Gama learned of mathematical navigation.

While al-Idrisi's accomplishments are impressive, this is not to overshadow the works of another brilliant Muslim geographer: al-Mas'udi. This 10th century genius from Baghdad traveled throughout the extent of the immense Islamic Empire, reaching such distant lands as southern Africa, India, Ceylon, Korea, and China. In his famous book, *Golden Pastures*, he thoroughly described the nature of the countries he had seen, "their mountains, their seas, their realms, their dynasties as well as the beliefs and customs of their inhabitants."[40] His book was the first of its kind produced, and its publication helped broaden Europeans' views of the Indian, Asian, and African continents.

Al-Mas'udi's writings were far more comprehensive than those of Marco Polo. He wrote and edited dozens of volumes on geography and history. While the likes of both al-Mas'udi (l0th century) and al-Idrisi (12th century) earned the admiration of the Sultans and commanded an entourage of pupils, Marco Polo (13th century), after returning from his trip to China, was denigrated in Europe as a mad man and fable teller.41

It was al-Muqaddasi who was perhaps Islam's greatest historical geographer. In the 10th century he produced such an impressive system of world geography that Sprenger coined him "the greatest geographer of all time."42 Dunlop notes that al-Muqaddasi relied exclusively upon the scientific method and that his work was exceptionally careful and rigorous. As perhaps his most impressive feat al-Muqaddasi, the brilliant systematizer that he was, created a technical method that is the very basis of geographical science today.

Yet, in the field of human geography even the famed al-Idrisi, al-Mas'udi, and al-Muqaddasi cannot compare to Islam's most learned and traveled geographer: Ibn Battutah. In 1325 at only 23 years of age he embarked on his wonderful series of journeys. His fantastic episodes of travel spanned some 30 years and encompassed a much wider range than that covered by Marco Polo. He gave the first account of the Maldive Islands, which are located off the southern coast of India and which he proclaimed "are one of the wonders of the world and number two thousand in all."43 Additionally, Ibn Battutah visited and made chronicles of Turkey, Arabia, Yemen, Afghanistan, China, India, Ceylon, Bengal, Malaysia, Indonesia, Spain, Mali, Niger, Nigeria, Sudan, Chad, Egypt, Algeria, Libya, Ethiopia, Tanzania, Mozambique, Madagascar, South Africa, Syria, Palestine, Iraq, Kurdistan, and Iran. His description of the city of Granada in Islamic Spain is particularly worth mentioning: "Its environs have not their equal in any country in the world (he had been to them all). Around it on every side are orchards, gardens, flowery meads, noble buildings, and vineyards."44

Ibn Battutah was the first explorer to reach remote regions of the Arabian peninsula as well as Western Sudan. Says Percy Sykes of Ibn Battutah's chronicles regarding these various regions, "the descriptions...are unsurpassed."[6]

In the Islamic Empire the cataloguing of the world was done in a systematic fashion and represented the combined work of innumerable scholars. Muslim geographers, of which there were thousands, produced the most accurate almanacs that had ever existed. These almanacs revealed incredibly useful and intriguing information regarding societies and lands previously unknown to the Western world. The first chronology of a trip from the West to China was accomplished in 880 A.D. by Abu Zayd, some 400 years before Marco Polo. In the southern Chinese town of Canton the Muslims established a seaport and colony as early as the 8th century. Al-Mas'udi reached the eastern edge of China during the 10th century and entered the South China Sea.[45] This was chronicled in his almanac. The term itself is derived from the Arabic, *al-manakh*.

Perhaps the most astounding achievement of the Islamic geographers was the development of the magnetic needle, which led to their invention of the compass. This discovery had such a profound impact upon the civilized world that its scope cannot be fully measured or described. They went further to perfect the use of the compass, sextant, quadrant, and astrolabe in navigation.[46] These devices were used by the Portuguese, Dutch, English, and Spanish explorers, whose accomplishments are taught as part of the routine geography classes in today's public schools.[47,48] However, no mention is made of the immense Muslim contribution.

In history classes throughout the Western world full credit for the discovery of the world at large, the Americas, Africa, India,

---

[6] For a contemporary Western view of Ibn Battutah's travels, see Abercrombe's feature article in *The National Geographic*, December, 1991.

China, and Japan, is given to the European merchants of the 15th through 17th centuries. In geography classes grade and high schoolers throughout the world are taught that Columbus was the first person since ancient Grecian times to propose that the earth was round and was the original Westerner to set foot upon the New World. Marco Polo is portrayed as the discoverer of China, while Vasco da Gama is regarded as the first non-Greek to discover India and East Africa. Additionally, the credit for final proof of the earth's sphericity is attributed to Magellan, the first navigator to sail around the globe. Thus, the obvious impression is that humankind was introduced to the unknown worlds solely by European pioneers. The contributions made by explorers of races or religions other than white Anglo-Saxon or Christian are entirely disregarded.

This wholesale omission of the contribution of Islam is reprehensible. Islamic innovations of the compass, sextant, quadrant, and astrolabe, as well as star charts, maps, and almanacs, made for the first time the achievement of long range navigation a reality. Now the world's captains, or, as the Muslims knew them, *umara al-bahr* (Princes of the Seas), could venture into the unknown waters and utilize sophisticated methods for maintaining a deliberate course. For the first time in history scientific navigation equipped captains with the capacity to know in advance how to reach their destinations and return safely. Thus, without these advances shipping would have been relegated to its ancient status of "staying to the shoreline."

The Islamic Era of the 8th through 14th centuries produced the first scientific navigators. Thousands of Muslim merchant and discovery ships plied the waters of the Mediterranean, African coasts, Indian oceans, South Seas, and seas of China. As a result of this scientific navigation, an international merchant trade developed such as had never before occurred in history. Trade was extensive not only in the Orient but also in the Christian West. This is evidenced by the wide range of navigational terms which have become incorporated into the English language, including the term for admiral itself—*ameer al-bahr*. Other Arabic derived words include sloop, cable, carrack, albatross, average (for calculating fees for goods), barge, felucca (a type of vessel), zenith (regarding the night skies), barge, monsoon, bark (a sailing ship), barkentine (a 3 masted vessel), and almanac. This vocabulary, denoting words still found in today's dictionaries, illustrates the extensive Islamic influence upon European navigation.

Prior to and during the early phases of the Age of Discovery the Europeans produced no works of significance on geography and navigation. The only Western works available were Ptolemy's *Almagest* and *Geographa*, which were archaic. Even these were not readily available until the latter part of the 15th century. Thus, Europe's navigators relied primarily on the Arabic/Islamic works. The 15th century voyage of Vasco da Gama is an example. As described by Boorstin after rounding the Cape of Good Hope da Gama secured a Muslim navigator named Ibn Majid, who guided him to his destination in Southwestern India. Ibn Majid was the author of the greatest sea-faring guide of its time: *Kitab al-Fawaʻid* (Nautical Directory). It appears he was well aware of al-Biruni's discovery made many years prior that the Indian and Atlantic oceans were connected.49

The Muslims of the Middle Ages were themselves outstanding navigators. They controlled the world's shipping industry for over five hundred years. They developed the lateen sail, which they originally used in the Indian ocean and then introduced to the Mediterranean. There, they upgraded the sail for application to larger ships. Lateen sails offered the unique advantage of driving ships in the desired direction even with unfavorable winds. This sail, with slight modifications, was the one adopted for the Spanish galleons, including Columbus' ships, the Nina, the Pinta, and the Santa Maria.50

The fact is Columbus mentions in one of his letters from Haiti that writings by Islamic Spain's Ibn Rushd (known to him as *Aventuez*), who had carefully documented the details of Muslim excursions into the Atlantic, caused him to guess at the existence of the Americas.51 Thus, the Muslims clearly initiated the discovery of the New World.52-55

As early as the 10th century Muslim navigators became aware of the wind patterns which determined shipping in that menacing direction: in other words, they knew about the Gulf Stream. Evidence is accumulating that their awareness of these winds caused them to venture towards the New World centuries prior to Columbus. However, no definitive statement can yet be made in this regard. What is certain is that they had no fear of sailing or, for that matter, crossing the Atlantic. They knew the world was round. Nor were they intimidated by the fear of the six-headed monsters and dragons that supposedly inhabited the dark Atlantic. This is confirmed by the fact that they explored the Azores, a group of islands located some 1,200 kilometers off the shores of Portugal.56 That amounts to a trip of nearly one-third the distance to the New World.

Muslim geographers were completely aware of roundness of the earth and that it, in fact, is in motion. This is crucial, because during this era the scholars of Europe believed the earth to be flat and motionless. Ibn Rasta, a 10th century Muslim geographer, said, "God be He praised, made the (celestial bodies) round as a ball....The earth is also round...and solid..." In providing evidence for this he stated, "the sun, moon and other planets do not rise and set...all over the earth at one time. They rise on its Eastern parts before they set on the Western parts."57 Further, Ibn Rasta noted that the different time periods which people across the globe view a lunar eclipse was evidence for the moon's orbit about the earth. During the 10th century al-Mas'udi wrote that "as the sun goes down over the oceans (of the West), it rises in the furthest end of

China, that being a revolution over one half of the earth's surface." Thus, he knew that the earth rotates about its axis.

To the Muslims of the 8th through 14th centuries the earth was literally a "geographical ball," its surface to be mapped, its features to be analyzed, its grandeur to be admired, and its inhabitants to be studied. All geographers agree that the beginning of geography, the catalyst for its projection into the realm of science, is the firm belief in the earth's roundness. This the Muslims knew. Yet, to know it was not enough. They proved it, astronomically, mathematically, and geographically. What's more, they infused that data within the Western world. Thus, the Muslims, the inventors of the globe, the encyclopedia, the atlas, and the almanac, are indisputably and solely the founders of geography as we know it today.

References

(see Bibliography for more detailed information)

1. Pasha, S. H., Ph.D. 1978. Personal communication.

2. Bammate, H. *Muslim Contribution to Civilization.*

3. *Dictionary of Scientific Biography.*

4. *Dictionary of Scientific Biography.*

5. Ibid.

6. Sarton, George. *Introduction to the History of Science.*

7. Asad, M. *The Message of the Qur'an.*

8. Mansfield, P. *The Arab World.*

9. al-Akkad, A. M. *Arabs Impact on European Civilization.*

10. Asad, M. *The Message of the Qur'an.*

11. Durant, W. *Age of Faith.*

12. Bammate, H.

13. Durant, Will.

14. Sykes, Percy. *A History of Exploration.*

15. *Dictionary of Scientific Biography.*

16. Durant, Will.

17. Ronan, C. A. *Science: Its History and Development Among the World's Culture.*

18. Pasha, S. H. Personal communication.

19. Bammate, H.

20. Ronan, C. A.

21. Sarton, G.

22. Nasr, S. H. *Islamic Science.*

23. *World Book Encyclopedia*, Vol. 1.

24. Sykes, Percy.

25. Akkad, A. M.

26. Durant, Will.

27. Sykes, Percy.

28. Bammate, H.

29. Durant, Will.

30. Sarton, George.

31. al-Akkad, A. M.

32. Bammate, H.

33. Dunlop, D. M. *Arabic Civilization to AD 1500.*

34. al-Akkad, A. M.

35. Dunlop, D. M.

36. Bammate, H.

37. Welty, P. T. *Human Expression: A History of the World.*

38. Bammate, H.

39. Dunlop, D. M. *Arabic Civilization to AD 1500.*

40. Bammate, H.

41. Welty, P. T.

42. Dunlop, D. M.

43. Sykes, Percy.

44. Ibid.

45. Ibid.

46. Welty, P. T. *Human Expression: A History of the World.*

47. Ibid.

48. Sarton, G.

49. Boorstin, Daniel. *The Discovers.*

50. Hayes, J. R. *The Genius of Arab Civilization.*

51. Akkad, A. M.

52. Ibid.

53. Dunlop, D. M.

54. Sykes, Percy.

55. *Dictionary of Scientific Biography.*

56. Akkad, A. M.

57. Ibid.

CHAPTER 6

Islamic Spain

The role of Islam in the creation of civilization is perhaps most readily appreciated by paying special tribute to one of the grandest eras of all: Islamic Spain. The Muslims entered Spain in the 8th century. They found it inhabited by primitive societies, who were culturally and intellectually backward. Their arrival was welcomed by the populous, which was suffering direly under repressive rule. Within 30 years they noticeably renovated the society. Within a century they transformed it into one of the most grand, advanced civilizations ever to inhabit the earth. Its great cities, Cordova, Seville, Toledo, and Granada, became the centers for the dissemination of knowledge and the advancement of civilization for five complete centuries.

The cities of Islamic Spain were international centers for cultural advancement as well as enterprise. What's more, they were the models for the urbanization of Europe, which was archaic. According to Goldstein the Islamic cities were "urban, commercial, sophisticated, exotic, and cosmopolitan." The fact is the Muslims created the world's first sophisticated, advanced civilization.[1-5]

Tenth century Muslim Cordova was a huge city with nearly one million inhabitants. With its paved streets complete with street lamps, 70 public libraries, numerous universities, and 800 public baths it was according to Peter Mansfield, "the most splendid city on the continent." The significance of this in respect to the development of human civilization is underscored by the fact that during that same period major cities in Europe were mere towns, inhabiting no more than 30,000 to 50,000 people at best estimates.[6]

As Europe's first metropolis Cordova quickly became the center of culture and learning in the West for people of all faiths. Scholars and clergymen came from all over Europe—Italy, France, Germany, Denmark, Holland, Switzerland, and England—to learn from the immense Islamic output. This environment is aptly described by Renan, who says there were "no barriers of race, culture, or nationality at the Muslim institutions. Muslims, Christians, and Jews studied on the same levels with complete racial tolerance." The fact is the Jews, who were tormented unmercifully by their former Christian rulers, flourished in tolerant Islamic Spain. Their glory under Islam was so great that this era is known as the Golden Years of Jewish science.[7] It was Islamic Spain which produced Musa ibn Maymun, known in Latin as *Maimonides*, the internationally renowned Jewish scientist. Ibn Maymun was the greatest Jewish philosopher and physician of all times.

As early as the 9th century Islamic Spain produced the original Western universities. Here, Islamic scholars and rulers organized and implemented elementary education as we know it today. In the rest of Europe organized education for the populous was unknown. In fact, it was never in place, not even in ancient Greece. Thus, the modern educational process owes its existence to the Muslims, who clearly were the first to institutionalize learning.[8-11] Preschools, grade schools, high schools, and universities were all their innovations. In Cordova alone 800 public schools operated, serving some 200,000 families. Within these schools and in the universities the modern concepts of varied curriculum, diplomas, licensure for professionals, and degrees were developed.[12-14] According to S. P. Scott education was so widespread that "it was difficult to find a . . . peasant who could not read or write."[15] At a time when the Kings of Europe were illiterate and could only sign their names with an "X," a Muslim ruler in Spain maintained a private library of some 600,000 books.[16]

Historically, the *madrasa* (Arabic for school) was the premier institution in Islam. Since the Prophet placed great emphasis upon both religious and scientific studies, Islam's conquerors, rulers, and scholars rapidly established the madrasa scholastic system in whatever land they controlled. This was the first institution installed after Salahiddeen's re-conquest of Jerusalem. He regarded it of greater import than the mosques. During the 7th century Caliph 'Umar ordered the construction of madrasas throughout the Islamic Empire. Unfortunately, the madrasa system has yet to be reestablished in the Middle East since the 12th through 16th centuries, when invaders burned these schools to the ground.

During the Middle Ages Spain was the only country in the European peninsula that disseminated knowledge. While Islamic Spain illuminated the world with wisdom, scientific achievement, and social sophistication, the rest of Europe systematically extinguished knowledge of any kind. There certainly existed "Dark Ages" but only in Europe east of its border with Spain and also extending throughout Asia all the way to the Sea of Japan.

Indeed, this was Europe's darkest era. Atrocities beyond comprehension were committed against the common man and scholar alike. If books of wisdom, science, or medicine were found, they were burned. Those who studied the sciences were forced to do so under the utmost secrecy. While the rulers of Britain burned at the stake anyone who claimed the sphericity of the earth, Muslim professors at the Spanish schools were teaching Christians and Jews geography by the globe.[17,18] Evidence exists that Columbus himself was a graduate of the Muslim institutions, where he reportedly received notions of the roundness of the earth.[19]

The Muslim rule in Spain was truly one of the greatest eras of civilization and science ever to occur in history. In contrast to the demagoguery in Medieval Europe Islamic Spain was governed by "justice, liberality, and refinement."[20] In what is an astounding achievement of social advancement this sophisticated civilization produced such modern refinements as public education, the public library system (complete with branch libraries), hot and cold running water, paper money, the concept of triple coinage, that is copper, silver, and gold coins, the police force, the banking system, high rise buildings, air conditioning, and the postal system, complete with air mail.[21]

There is a compelling reason for this tremendous improvement in civilization, considering that the rest of the world was immersed in barbarism. For the first time in history individuals of all faiths and cultures were treated with complete tolerance. Thus, the talents of dozens of civilizations were integrated under Islam. As a result, humanity worked in unison to create a civilization of such grandeur, of such height that it has never been duplicated even in modern times. The glorious Islamic rule lasted from the 8th to 15th centuries, a period of over 700 years. That is more than three times the period of time that the United States has been in existence.

Prior to Islamic rule Spain was steeped in a state of ignorance similar to that found in the rest of Europe. The Christian Kings had fully repressed intellectual and religious freedom. At the hands of the Muslims Spain was rapidly transformed into one of the most spectacular societies of its time, with possibly only the glory of Baghdad exceeding it. In Islamic Spain scientists operated freely, in fact, their efforts were vigorously encouraged by the State.[22] In Europe they were tortured, imprisoned, and killed.[23] Thus, it becomes evident that it was through its contact with Islamic Spain that Europe first learned of democracy, freedom of government, press, speech, and religion.[24-27] Yet, Europe's, as well as America's, greatest debt is owed to Muhammad, may God rest his soul. This one man initiated each of the revolutionary social advances which are the very crux of modern civilization and which were unknown before his time. There was no freedom of expression, democracy, or religion in any previous ancient society, not in Egypt, Greece, Persia, China, and certainly not in Rome. What's more, after Islam the 15th-17th century Spaniards, as well as the Portuguese, British, and Dutch, completely repressed freedom of religion and speech. This is precisely why the original immigrants landed in America, that is to avoid religious persecution. The fact is there is no evidence of democracy arising independently from European civilization. Had it not been for Muhammad, who prevailed in establishing these principles against enormous obstacles, it is reasonable to presume that democratic civilization as we know it today would fail to exist. Yet, the civilization initiated by Muhammad was in many respects even grander than the so-called democratic societies of today, because in an Islamic civilization the races are treated equally, that is racism of any sort was/is abolished. In other words, no one is a "foreigner."

Muslim Spain was the conduit through which knowledge from the Islamic Empire spread throughout the Eurasian continent.[29-31] Seville, Cordova, Toledo, and Granada all were centers of Islamic learning and culture. These cities were the models for Europe's democratization.[32-35] Christians and Jews came to them to learn civilization and science, mastering the Arabic language to do so. They attended the Muslim-constructed universities, learning medicine, pharmacology, botany, geology, geography, sociology, chemistry, physics, mathematics, astronomy, literature, and philosophy. Few people realize that for fully five centuries Arabic was the language to master for anyone who wished to learn the sciences.

Here is an interesting note. During the Middle Ages Pope Syllvester II (9th-10th centuries) was one of the few literate clergy. He received his education prior to his commissioning and that, of course, was at the Islamic universities in Spain. After his appointment he showed an interest in pursuing the science of medicine and so fell under the suspicion of sorcery. He escaped the witch-burners and lynchers during that precarious time only because of his high office.

With time Europe desired more direct contact with Islamic learning. Under the direction of the Muslims Constantine built Europe's first medical school, and a Muslim scholar was hired as its director. The Europeans contracted with Islamic scholars in order to institute public education within their countries. The establishment of the West's first astronomical observatory and public university is to the Muslims' credit. During the 10th century in Italy Muslim scholars organized the University of Salerno, which became Europe's premier medical school. Its textbooks and curriculum were entirely Islamic creations. F. H. Garrison claims that this school was the key one for propelling Europe's barbaric and archaic medical practice into true science.

The Decline

It was during this era of European awakening that the output of Muslim science began to decline. All societies rise and fall, and this Islamic one was no exception. However, the Islamic society remained the leader of the free world for a longer period than any other society prior or since. Yet, in regard to human interactions this was perhaps the most unique of all empires ancient or modern. It is the only example of a culture which after proclaiming to be the leader of the free world actually adhered to that philosophy.36

Throughout the course of its rule tolerance for peoples of other faiths, cultures, and colors was maintained. Even at its pinnacle of cultural and military power, no sign of racism could be found. Everyone, regardless of race or creed, had the opportunity to excel, to become scholar, judge, discoverer, or scientist. It was Islamic civilization which produced the scholarly al-Jahiz. A former black African slave he wrote over 200 books and has been announced as one of the greatest literary geniuses of all time.[37,38] In contrast, in the United States as late as the 1880s it was illegal for black Americans to learn to read and write, and those who became literate did so under the utmost secrecy.

Perhaps the most dominant characteristic of the Muslims of Spain was their thirst for knowledge. No previous culture created and disseminated such a vast amount of knowledge. Incredibly, hundreds of thousands of books could be found in a single man's library. Previous civilizations were not nearly so productive in the realms of the arts and sciences, not even the Greeks. In fact, the quality and quantity of the latter's works are a distant second compared to the vast productions of the Islamic scholars. This is aptly summarized by Goldstein, who notes that the scientific knowledge compiled by Islam was "without question the most complete store" ever produced in human history.

The Greeks ruled a minor empire of city-states. They assimilated ancient knowledge primarily from Phoenicia, Egypt, and Babylon. Islam owned a massive empire ranging from the edges of China, Russia, India, and Afghanistan to the Western border of France. As a result, the Muslims captured ancient knowledge from the whole of civilization: India, Persia, China, Africa, Samarkand, the Indies, Russia, Egypt, Rome, and Greece. They preserved the crucial knowledge accumulated by all of these societies and then updated it extensively. They created a practical science unknown to the Greeks. What's more, they organized this knowledge into a form which Europe could readily assimilate.

When the West overcame the Islamic civilization it did precisely the opposite. Western conquerors destroyed the books and essays of Islam by the millions, brutally killed its scientists, and burned to the ground its centers of scholarship.[39-42]

The collapse of the Islamic society was largely perpetuated by barbarians, who, in their rage, decimated anything of value. Through their aggression the Muslim centers of knowledge and culture were systematically eliminated from the globe. In Spain the Muslims were attacked by fanatics, such as the Vikings, Franks, and Crusaders, who plundered and pillaged, leveling to dust the world's greatest libraries, universities, laboratories, and observatories. These conquerors had no interest in advancing civilization. Rather, they raped it as a result of their own selfish desires. As a consequence Spain was returned to an ancient level of ignorance, from which it has never recovered.

The Islamic schools and libraries contained untold millions of priceless manuscripts, the Cordova library alone possessing some 400,000. Books were prized in this civilization much like fine works of art are in ours. Everyone who possessed money or status collected them. Hundreds of bookstores, libraries, and schools could be found in Cordova, Seville, Granada, and Toledo. This places the quantity of books in Islamic Spain alone at several million. Additionally, each mosque, of which there were thousands, contained an immense library. In some libraries hardly a book survived. The barbaric illiterate Franks who conquered Spain saw nothing useful in these masterpieces. Driven by hatred and ignorance, they burned books in massive bonfires by the millions.

During their conquest of Islamic Spain the Europeans slaughtered hundreds of thousands of innocent Muslim men, women, and children and burned many thousands of others at the stake. It was a bloodbath beyond comprehension. The entire population of Cordova, a city of some one million people, was exterminated. The few who survived were either tortured to death, exiled, or enslaved.

Ironically, the vast majority of these victims were indigenous Spaniards, who adopted Islam. These former Christians sought refuge in Islam from the tyranny of their previous rulers, the Visigoth Christian kings. In Granada, the major city closest to the African/Asian Islamic Empire, only 500 of a population of 200,000 had Arabic or African parents.[43]

As a result of these atrocities Spanish civilization received a monumental setback, one that repressed intellectual development for centuries—perhaps permanently. This incineration of the greatest, most productive centers of learning that had ever existed on European soil produced a backlash of such an immense scope that it is beyond comprehension. Certain notable historians claim that the continuation of this Islamic culture would have accelerated the Renaissance by some seven hundred years. This is an impressive and, yet, realistic statement. Taken literally it means that Medieval Europe would have been rapidly driven from barbarism into the technology of the early 20th century.

Still, the Renaissance is a direct consequence of the Islamic advances. Beginning in the 12th century the Islamic works in the sciences, as well as in literature and philosophy, were translated into Latin. This was at a time when Europe possessed few books, and libraries were essentially non-existent. The fact is there were no public libraries in all of Europe during this period. In contrast, Baghdad alone boasted thirty-six, Cairo, twenty, and Cordova, seventeen.

Yet, despite the illiteracy of Europe's masses the West did have its wise men. These individuals were well aware of the grandeur of Islamic Spain. They quickly recognized that Islam had preserved much regarding the advancement of civilization in its books, that is what was left of them. Thus, the intellectuals of Europe attached great importance to the books of Islam. With a passion unprecedented in European history they devoured the recently captured libraries, seeking the causes for the superior Islamic civilization. Thousands of Latin renditions were produced primarily in the 13th through 15th centuries. This mass translation was made possible for three major reasons.

First, even before the Renaissance hundreds of European scholars learned the Arabic language, as well as science, philosophy, music, and art, at the Islamic universities. These international universities were found primarily in Spain, but they also existed in Sicily, the former Islamic colony, southern Italy, and North Africa. Interestingly, a majority of these scholars were Englishmen.

Some Europeans stayed in Islamic Spain for decades, essentially passing their entire lives translating into Latin the Islamic works on the sciences, literature, and philosophy. Gerard of Cremona (d. 1187), who spent decades in Toledo and translated over 90 Arabic works into Latin, is just one example. His feats include the rendition of Ptolemy's Almagest and Ibn Sina's *Canon of Medicine*. Goldstein writes in his book, *Dawn of Modern Civilization*, that European scholars came to Spain to study and translate the Islamic books "in droves." Will Durant says that after 1150 A.D. these translations began to "pour into Europe" from Islam. As a result of this enormous effort Europe acquired virtually all of the existing works on the sciences, as well as literature and philosophy, prior to the destruction of Spain's Islamic civilization.

Secondly, the capacity for mass production of books was a direct result of the introduction of paper manufacturing to Europe from the Islamic world. Paper was in use at Mecca in 707 A.D. It was found in Spain in 950 A.D. but was not found in England until 1309 A.D. Baghdad operated the world's first industrial paper mill during the beginning of the 8th century. It was not until some four centuries afterward in the latter part of the 12th century that Europe entered the paper milling business. Before this time book production in Europe was unknown and the populous was largely illiterate. The introduction of paper manufacturing set the stage for the production of Latin translations on such a large scale that cultural upheaval was imminent. The library in Cairo alone offered two million books to facilitate the Europeans' translation efforts. The fact is there was no scientific literature available in Europe prior to Islam.

Thirdly, in the 11th century Christian armies captured certain cities in Spain, notably Toledo, virtually intact. Here the most massive translation movement ever in Christendom occurred. The Christians relied upon Spain's Jews, who were the major contributors to this effort, as they had adopted the Arabic language several centuries prior. Long before, the Muslim educators had been killed, imprisoned, exiled, or enslaved.44,45

During the Middle Ages Europeans lived in a society devoid of even the most rudimentary elements of civilization. Books, colleges, libraries, public education, and other social refinements were unknown. Thus, Europe relied heavily upon the civilization of Islamic Spain to gain a glimpse of culture. The scholarly and the unlettered alike came from all over Europe, as well as Asia, to gain from the immense stores of knowledge accumulated by the Spanish Muslims. By the end of the 13th century Islamic science, literature, philosophy, and art had been securely grafted within the minds of the learned men of Europe. These writings so strongly stimulated Western thought—the mathematics of al-Khwarizmi (L - *Algorismus*), the optics/physics of al-Haytham (L - *Alhazen*), the medical ethics and practice of ar-Razi (L - *Rhases*), the medicine and philosophy of Ibn Sina (L - *Avincenna*), the surgical and medical expertise of az-Zahrawi (L - *Abulcasis*), the mathematical and astronomical wisdom of al-Battani (L-*Albategnius*), and the philosophy and scholarship of Ibn Rushd (L - *Averroes*), that the West was forced to undergo a "Renaissance."

Prior to Islam European civilization was hopelessly mired in barbarism. Chained by religious dogma, it stood little chance of recovery. Only a massive intellectual transformation could dislodge it. This is precisely what occurred. For the first time in history the knowledge of an advanced civilization was transferred directly to an inferior one. Usually, such knowledge, whatever remains of it, is recovered centuries after the decline, when it is too late to benefit. In the case of the Islamic civilization the transfer was immediate and the information largely intact. Plus, the transfer occurred *while the Islamic civilization was at the height of intellectual achievement.* This is precisely why the names of Islamic scholars were far more renown in Europe than those of the ancient Greeks. The evidence of this influence is so dominating that Sarton, Durant, Goldstein, DeVaux, Hill, and Humboldt proclaim that modern science, as well as civilization, owes its existence to Islam.

After the loss of Islamic rule Spain's indigenous population failed to cultivate the sciences. It is true that the Christian scholars of Spain, as well as its clergy, studied the Latin translations of Arabic works in great depth. They frequently referred to them in their writings. Yet, they entirely failed to perform original research for furthering the cause of science. Instead, the Spaniards of the 15th through 18th centuries mastered the practices of witchcraft, superstition, astrology, and sorcery.

What they did extract from the Islamic works were any methods by which they could fill the coffers of the King's treasury as well as their own. They performed the latter with great effectiveness. One can only imagine how this might have developed; "O King, the world is round, not flat as we have presumed. Let us seek treasures and riches for you, for the Church and for God. Prepare for us a fleet so we may discover a new avenue to reach the Spice Islands." Or, "O King, we have just learned that the passage to India can be made safely; the Arabs say the seas of India and Africa connect. Let us proceed for whatever riches we might find." However, the correct verbiage is, in reality, plunder, not find.

What, then, was the heritage left to the world by the Spaniards, considering that the shelves of original contributions in the sciences remained bare? After Islam Spain led the world in violence, tyranny, repression, colonization, and slavery. It was the Spaniards who, along with the Portuguese and English, initiated that gruesome stage in human history: the black African slave trade. Thus, the Spanish were thieves and pirates, plundering and pillaging every society, culture, and land they possibly could for some 200 years until their rapid decline. During their rule they failed to contributed to society on any significant scale, nor have they contributed since. Rather, they utterly destroyed the civilizations they encountered as well as their own. The Spaniards exterminated races. Islam preserved, in fact, cultivated them. Thus, when Spanish Islam was incinerated, so disappeared any degree of tolerance and sophistication. The Spaniards' call to power was one of the shortest of any dynasty, spanning little more than a century.

The Coming of the Mongols

The barbarians of the Middle Ages were not content with the destruction of Spain. The Crusaders continuously invaded the Fertile Crescent and were responsible for the unprovoked slaughter of tens of thousands of individuals. After 200 years of debauchery they were ultimately driven off. The Crusaders came not to advance culture and civilization but to destroy it. However, they were followed by the greatest butchers of all time. In the thirteenth century hordes of Mongols led by Genghis Khan attacked and decimated Islamic Persia and, along with it, its great centers of scholarship. A weakened Islamic Empire was easy prey for invaders, who repeatedly harassed the Fertile Crescent, further crumbling it. Ultimately, the gruesome fanatics from Asia entered Baghdad, the core of Muslim learning. The Mongol hordes sacked the defenseless city, indiscriminately massacring men, women, and children by the thousands. The masses were ruthlessly obliterated; the clergy in the mosques, the patients, nurses, and doctors in the hospitals, and the professors and students in the schools, all were put to the sword. The world's greatest institutions of science and medicine were looted, pillaged, and burned to the ground. On their forty day rampage of Baghdad the Mongols systematically destroyed the most advanced collection of human knowledge of all times. Tens of millions of books were lost forever. In what were acclaimed as the most advanced libraries that have ever existed barely one in a thousand books escaped historical doom. Not a trace remained of Baghdad's internationally renowned House of Wisdom (*Bayt al-Hikma*), which was then regarded as the world's most imminent teaching center. Seven centuries of labor, the accumulated knowledge of all of the ancient civilizations, plus the prolific writings of tens of thousands of Islamic scholars, all were destroyed in minutes.[46-50]

It seems incomprehensible that an entire civilization could have been erased from this earth in a matter of hours, yet, indeed, that is precisely what happened. The most enlightened city on the earth, Baghdad, was crushed into oblivion, never again to regain its grandeur. According to Durant this was the first time in history that a complete civilization was abolished so quickly. It was a destruction which repressed the development of civilization for centuries. What an enormous wastage of human resources and what a dark day in human history that was.

The extent of the havoc wreaked is best comprehended by reviewing eyewitness accounts. The witnesses reported that the rivers flowing through Baghdad were so saturated with blood and books that they turned black. The streets, it is said, were filled ankle deep with blood. This city was inhabited by hundreds of thousands of people: few survived. Pregnant women and women holding their infants were cut down, stabbed, and beheaded and their youngsters killed. Yet another account claims that the Tigris river was dyed red for several miles along its course as it was fed by the blood of six horrifying days of slaughter and butchery. [51]

All told, some 800,000 people perished in Baghdad alone, amounting to over 90% of its population. Each of its universities, observatories, hospitals, public libraries, and nearly all of its 3,000 mosques were completely destroyed. Ibn Battutah visited the region some 100 years later, noting that it was still "largely in ruins."

Ironically, the same people who orchestrated these crimes later converted to Islam, the very thing they attempted to destroy.51 Hulagu, the grandson of the chief Mongol barbarian, eventually constructed an astronomical observatory in Baghdad similar to the ones his people destroyed. The conquerors found Islam to be a more sensible, tolerant, refined, and, certainly, more "civilized" method of life than any other religion they might have previously encountered or adopted. Perhaps this is because Islam, rather than being a religion, is a system of rules and regulations.52 Its purpose: the creation of elegance in civilization.

As a result of these senseless crimes all of humanity suffered a reversal, the extent of which cannot be measured. The devastation was so thorough that it completely eliminated Islamic civilization—rather, global civilization—in that era. In fact, Islamic civilization was so deeply buried that never again did it contend as the world leader of science, civilization, and culture. Certainly, there was a degree of revival during the rule of the Ottomans as well as the Monguls of India. Yet, never again did such a civilization of the grandeur of Islamic Spain, Baghdad, Damascus, or Persia ever surface in these regions again.

Yet, the full blame should not be placed on the conquerors. The Muslims themselves perpetuated their sorry state. Islamic historians correlate the decline of Muslim science and culture to a negligence of their own making: the failure to carefully adhere to the principles of the Qur'an and the teachings of the Prophet Muhammad. All of the values necessary for the advancement of society, morally, culturally, and scientifically, may be found therein. These values were firmly established in civilization by Muhammad, may God rest his soul. The conservation of resources, public sanitation, moderation in diet, abstinence from alcohol, free trade, religious tolerance, common decency, equality of the races, and freedom of speech were all his innovations. However, with time these values were ignored, forgotten, perhaps defied. During the Islamic Era pomp and gluttony depleted resources, weakening civilization economically, physically, and militarily. Dietary abuses, sugar addiction, alcoholism, and even homosexuality riddled society. Schisms, that is internal political strife, as well as greed, divided and weakened the Caliphate. The region was ripe for conquest. Ironically, the same factors have fractured the so-called Islamic world of today. The fact is, since Islam is defined herein as elegance in civilization, the term "Islamic World," currently in vogue for the Middle East and other regions inhabited by Muslims, is invalid. There has been no Islamic world since the fall of Granada in 1492 A.D. Furthermore, feudalism, tribalism, racism, anarchy, poverty, and, particularly, illiteracy, are rife within the lands of predominately Muslim population, the wealthy Sheikdoms being the epitome of this dilemma. All of these aberrations are the antithesis of Islam. Greed and anarchy rule but not Islam, at least, not yet.

Concerns of the West about the revival of Islamic civilization are another reason it has yet to resurface. The Western powers are afraid of it. Fear leads to repression. Coups are performed, CIA stooges are installed, antagonistic literature is disseminated, and propaganda is unleashed. Thus, Western powers unfurl a barrage of tactics to combat and/or prevent the onslaught of Islamic civilization. What's more, they spend tens of billions of dollars in the process.

A fear of Islam is ludicrous. Furthermore, fear of Islamic revival is particularly bizarre. Islam can't be revived. It already exists. Islam created civilization. Thus, even if it were "revived" it would merely advance civilization further. Sarton made it clear that if Islamic Spain were left intact instead of being destroyed, civilization would have been catapulted by some 700 years. However, to state that Islam "created" civilization is somewhat misleading. This is because Islam continues to create civilization. In fact, it exists with us today. It thrives amongst us in the form of the most prominent modern institutions: the postal service, the monetary unit, the bank, public education, the university, the observatory, and the library. It advances us with mathematics, physics, medicine, philosophy, ethics, liberality, and morality. It refines us with sophistication, manners, and culture. It protects and beautifies us with soft cotton, fine leather, fur lined boots, and patterned silk. It cleanses us with running water, the hand towel, the handkerchief, the table napkin, the toothbrush, soap, deodorant, and toiletries. It nourishes us with spinach, asparagus, peaches, watermelon, oranges, grapefruit, lemons, bananas, rice, strawberries, apricots, fruit juice, herbal teas, ginger, cinnamon, spices, and almonds. It relaxes us with soft music, floral gardens, water fountains, fine bedding, and the sofa. It defines us with intellect, justice, and the moral code. It medicates us with essential oils, herbal remedies, tinctures, ointments, flower essences, and the balanced diet. It educates us with encyclopedias, atlases, almanacs, globes, textbooks, journals, and the university. Thus, if an individual is distressed about Islamic revival, his/her fear is particularly bizarre, because this is how we live today.

Yet, Islam, as a political/governmental system, fails to exist today. What's more, the fragments of it which do exist are not found in the so-called Muslim countries. Rather, the remnants of Islam are found in Western civilization in the form of law and order, fine goods, commerce, free trade, and social sophistication. In fact, the United States, with its high rate of literacy, freedom of speech, public education, public sanitation, law-based civilization, and organized social structure, exhibits a greater degree of "Islam" than any of the various countries of "Muslim" inhabitants. After all, it was Islam which provoked the West to leap from social degradation to the realm of advanced civilization, a fact which has been thoroughly proven by the writings of a wide range of historians, including Goldstein, Sarton, Durant, Humbolt, Renan, Bernal, Hill, Turner, and countless others. What's more, as a result of the enormous revolution in the production of scientific data Islam's influence upon Western civilization is continuous and pervasive: the establishment of the Arabic numerals, upon which all computerization depends, is merely one example. Ironically, the United States and other Western countries reap uncountable benefits from Islam, while waging an aggressive campaign to prevent it from benefiting the people who need it the most: those in primitive societies as well as the impoverished Third World.

Islam had to be responsible for the advancement of civilization. It commands that universal philosophy which fertilizes humanity no matter what is its condition. It does so by melding together all of the talents of the world population. It achieves it by bringing the best out of the individual, morally, physically, intellectually, and scientifically. Middle Age Christendom had no such record. In fact, it did precisely the opposite. Christian rulers practiced intolerance with a vengeance, deliberately obstructing the advancement of civilization. In the epitome of ignorance organized religion systematically tortured and murdered countless thousands, whose only crimes were to seek the advancement of science. What few scholars existed studied in hiding and absolute fear. Roger Bacon, infamously known in the 13th century as "Doctor Mirabilis," was compelled to leave the study of science and philosophy. He was tabbed by his Western contemporaries as a satanic dabbler. Mobs chided him, yelling for "this sorcerer's hand to be cut off" and "this Muslim to be exiled."

Even the Renaissance failed to halt the persecution. Galileo, the great Italian astronomer, was forced by the Church to deny his Copernican based theory that the earth revolves around the sun as satanical heresy. Bruno, one of Europe's most imminent scholars, was mercilessly burned at the stake for his attempts to advance knowledge. While the Islamic clergy and politicians fully encouraged the development of the sciences, the Western rulers and theologians felt obliged to continuously repress them. The sciences would certainly have failed to arise under such intense pressure. Without intervention there was little hope for their recovery. That intervention came from Islam.

The accomplishments of the Muslim scholars and scientists of the Middle ages are vast. Their influence is persistent, because much of their work is essentially modern. In other words, many of the educational methods in place today—in mathematics, physics, geography, geology, chemistry, and medicine—are the same as they were in the Islamic Empire some 1000 years ago. Yet, only a modicum of the Islamic scholars' accomplishments have been analyzed. The majority of their works remain unpublished.

Muslim scholars were scientists in the purest sense, using logic and reason and then proving their ideas with scientific experiments. As a result, they were the most advanced chemists, botanists, physicians, druggists, biologists, mathematicians, engineers, surgeons, veterinarians, physicists, geologists, geographers, astronomers, sociologists, and philosophers of their time. In the field of chemistry the Western world produced none of their calibre until the advent of Boyle and Lavoisier in the 17th and 18th centuries. In astronomy they had no European contemporaries until Kepler and Galileo, both of whom were influenced by them. Isaac Newton was the first to exceed their accomplishments in physics, some six hundred years after al-Haytham and al-Biruni. In mathematics they had no equals until the 19th century, and even today many of their accomplishments remain untouchable. In human geography even today's best scholars are their inferiors. Pharmacologists of their expertise have only existed since the 20th century. In herbal pharmacology they have yet to be exceeded. In medicine currently they remain unrivaled. Ibn Sina, ar-Razi, Ibn Zuhr, Ibn an-Nafis, az-Zahrawi: which of today's physicians could match their ingenuity, scholarship, diagnostic wisdom, and literary output? Even in the arena of modern medicine's pride, diagnostic acumen, today's physicians would be their subordinates.

The astounding accomplishments of the scholars of Islam created massive change, propelling humankind from anarchy, illiteracy, and barbarism into the realm of modern civilization. This was a modern civilization, which advanced humanity without damaging the environment or discriminating against other cultures. That Islam induced a profound, positive, original revolution in humanity is described in great detail by the world's top Western historians. Briffault aptly summarized this momentous achievement; "Nowhere is...the decisive influence of Islamic culture so clear as in the supreme source of its victory: natural science and the scientific spirit." Bernal states that the Muslims may be regarded as the sole founders of modern chemistry. Yet, it was Goldstein who perhaps best edified the enormity of the Islamic achievement by proclaiming that every Western science owes its origins to Islam, which is confirmed by Humboldt and Sarton, who write that the Muslims were the founders of all of the basic sciences.

References

(see Bibliography for more detailed information)

1. Goldstein, T. *Dawn of Modern Science.*

2. Durant, Will. *Age of Faith.*

3. *Dictionary of Scientific Biography.*

4. Hill, D. R. *Islamic Science and Engineering.*

5. Lane-Poole, S. *Story of the Moors in Spain.*

6. al-Akkad, A. M. *The Arab's Impact on European Civilization.*

7. Goldstein, Thomas. *Dawn of Modern Science.*

8. Pasha, S. H. Personal Communication.

9. Sarton, George. *An Introduction to the History of Science.*

10. Durant, W.

11. Burk, J. *The Day the Universe Changed.*

12. Sarton, G.

13. Durant, W.

14. Pasha, S. H.

15. Sarton, G.

16. Durant, W.

17. Turner, H. R. *Science in Medieval Islam.*

18. Sarton, G.

19. Durant, W. and *World Book Encyclopedia.*

20. Durant, W.

21. Ahsan, M. M. *Social Life Under the Abbasids.*

22. Durant, W.

23. Durant, W. and Pasha, S. H.

24. Goldstein, T.

25. Sarton, G.

26. Durant, W.

27. Sarton, G. and Pasha, S. H.

28. Sarton, G. and Durant, W.

29. Durant, W. and Sarton, G.

30. Goldstein, T.

31. Singer, Charles. *A Short History of Scientific Ideas.*

32. Durant, W.

33. *Dictionary of Scientific Biography.*

34. Dunlop, D. M. *Arab Civilization to A.D. 1500.*

35. Pasha, S. H.

36. Durant, W. and Pasha, S. H.

37. Dunlop, D. M. see also Ahsan, M. M. *Social Life under the Abbasids.* (pp. 2-5)

38. *Dictionary of Scientific Biography.*

39. Sarton, G.

40. Durant, W.

41. Pasha, S. H.

42. Burk, J. *The Day the Universe Changed.*

43. Irving, T. B. *The End of Islamic Spain*

44. Goldstein, T. and Pasha, S. H.

45. Garrison, F. H. *An Introduction to the History of Medicine.*

46. Singer, Charles.

47. Garrison, F. H.

48. Burk, J.

49. Durant, W. and Sarton, G.

50. Pasha, S. H. Community lectures.

51. Ibid.

52. Ibid.

Appendix A

List of Islamic Scholars

It would be impossible to produce a comprehensive list of the scientific scholars of the Islamic Era. This period was nearly eight centuries long. What's more, unfortunately, most of the writings of the Islamic scientists have been lost. Even so, what remains is in itself a wealth of knowledge. Few if any of their writings and books have been translated into modern English. The information produced by these scholars is of such enormity and complexity that it would take decades, perhaps centuries, to elaborate. As originally described by S. H. Pasha of all previous civilizations it was the Islamic civilization alone which produced diversified scholars, that is scholars capable of publishing in numerous fields. Certain scholars, such as al-Biruni and Ibn Sina, were published in virtually all of the sciences. The following list is an attempt to provide a partial guide of the most well known of the Islamic scientists, men whose writings are still "alive," either in Arabic or Latin texts or as English versions.

Arabic Name and Century	Subjects of Expertise
Muhammad bin Ahmad (10)	mathematics
	astronomy
Ibn Bajjah (11, 12)	philosophy
	physics
	mathematics
	philosophy
	medicine
	botany

al-Battani (10)	astronomy mathematics physics trigonometry
Ibn Batuttah (14)	jurisprudence geography
Ibn Baytar (14)	pharmacology botany medicine
Ulug Beg (14, 15)	astronomy mathematics geography
al-Biruni (10, 11)	astronomy physics geology geography mathematics botany pharmacology medicine biology obstetrics mineralogy
Ibn Butlan (11)	public health nutrition
Kamal ad-Din (13)	optics physics mathematics zoology astronomy
ad-Dinawari (9)	medicine botany
al-Farghani (9)	pharmacology astronomy mathematics

al-Farabi (10)	astronomy physics philosophy zoology
Ibn Firnas (9)	optics mechanics engineering physics mathematics
al-Ghafiqi (12)	surgery ophthalmology
al-Harrani (9, 10)	astronomy mathematics medicine
Abu'l Hasan (10)	astronomy physics mathematics
Habash al-Hasib (9)	astronomy mathematics
al-Haytham (10, 11)	optics mathematics physics astronomy trigonometry engineering meteorology
al-Idrisi (12)	geography geology botany
'Ali ibn 'Isa (10, 11)	medicine botany pharmacology ophthalmology

al-Jabr (10)	chemistry botany pharmacology medicine
al-Jazari (12, 13)	physics mechanics engineering mathematics
Ibn al-Jazzar (10)	medicine public health pediatrics
Ibn Juljul (11)	botany pharmacology medicine
al-Kashi (14, 15)	mathematics astronomy medicine geology
Ibn Khaldun (14, 15)	philosophy mathematics
al-Kharaji (10, 11)	physics astronomy mathematics
al-Khazini (11, 12)	mathematics astronomy physics engineering
'Umar Khayyam (11, 12)	mathematics physics astronomy
Ibn Khirdazbah (10)	astronomy geology geography

al-Khwarizmi (8, 9)	mathematics astronomy geography geology
al-Kindi (9)	medicine astronomy optics botany chemistry philosophy mathematics pharmacology medicine physics biology zoology
al-Maghribi (13)	astronomy mathematics
al-Mahani (9)	astronomy mathematics
al-Mahusin (13)	surgery ophthalmology pharmacology
al-Majusi (10)	anatomy medicine botany philosophy
al-Mas'udi (10)	geology mathematics geography
al-Muqaddasi (10)	geography geology mineralogy meteorology

Ibn an-Nafis (13)	anatomy physiology medicine surgery
Ibn Rasta (9)	astronomy geography geology
ar-Razi (9, 10)	medicine psychology physiology botany pharmacology medical ethics chemistry biology
Ibn Rushd (12)	philosophy jurisprudence medicine botany physiology public health nutrition
Ibn al-Quff (13)	surgery anatomy medicine physiology embryology
Thabit bin Qurrah (9, 10)	astronomy mathematics physics botany pharmacology

Ibn Sahl (9)	embryology pediatrics gynecology obstetrics
Ibn Sarabiyun (9)	medicine pharmacology botany
Ibn Sina (11, 12)	medicine philosophy astronomy botany pharmacology geology biology chemistry physiology pathology mineralogy geography zoology
Ibn Tufayl (10)	philosophy
at-Tusi (12, 13)	astronomy mathematics physics geology philosophy mineralogy
al-'Urdi (14)	mathematics astronomy
Ibn Abi Usaybi'ah (12, 13)	medicine medical history
Abu'l Wafa' (9)	mathematics astronomy physics

Yaqut (13)	geography geology botany zoology
Ibn Yunus (10, 11)	physics mathematics astronomy
az-Zahrawi (11, 12)	surgery anatomy medicine pharmacology anesthesia obstetrics botany dietetics
Ibn Zuhr (12)	medicine & surgery botany anatomy pharmacology pathology physiology dietetics

Table Two

Words of Arabic Origin

Arabic words entered the English language through a variety of channels. Many of these words spread into Europe from Islamic Spain as a consequence of the translation of Islamic books into Latin. An early version of Webster's dictionary lists Arabic as one of the top five languages from which English is derived. In the 1930s Walt Taylor recorded approximately 1,000 English words of Arabic origin, and this list would grow to many thousands if their derivatives were included. Only a small number of them are listed here. The primary source for these words is *Webster's International Dictionary* (large edition, 1918). Most of these words were incorporated into the English language many centuries ago. Where possible, the Arabic word from which it is derived is included.

abutilon (*aubutilun*)

acerola (*az-zu oor*)

Achernar (*akhir nahr*)

admiral (*amir al-bahr*)

albacore (*al-bakurah*)

albatross (*al-qadus*)

alcalde (*al-qadi*)

alcazar (*al-qasr*)

alcohol (*al-quhul*)

Alcor

alcove (*al-qubba*)

Aldebaran (*al-dabaran*)

aldehyde

alembic (*al-anbiq*)

alfalfa (*al-fisfisah*)

alfilaria (*al-khilal*)

alforga (*al-khuri*)

algarroba (*al-kharruba*)

algebra (*al-jabr*)

Algol (*al-ghul*)

algorism (*al-khwarizmi*)

algorithm (*al-khwarizmi*)

alidade (*al-idada*)

alizarin (*'acarah*)

alkali (*al-kili*)

alkanet (*al-hinna*)

Alkes (*qa'idat al-batiya*)

Almagest (*al-majisti*)

almanac (*al-manakh*)

almucanter (*al-muqantaral*)

aloe (*allueh*)

Alphecca (*nayyir al-fakka*)

Alpheratz (*surrat al-faras*)
Altair (*al-nasr al-ta´ir*)
Altaref
aludel (*al-uthal*)
amber (*anbar*)
ambergris (*anbar*)
aniline (*an-nil*)
antimony (*antimun*)
apricot (*al-birquq*)
arabesque (*Arab*)
Arkab (*Arkub*)
arsenal (*dar sina' ah*)
artichoke (*al-kurshuf*)
assassin (*hashshashin*)
attar or attar of roses (*utur*)
aubergine (*al-bahinjan*)
average (*awariyah*)
azedarach (*azadirakht*)
azimuth (*as-sumut*)
azoth (*az-za uq*)
azure (*lazaward)*
azurite (*lazaward*)
barberry (*barbaris*)
barge
benzene (luban jawi)
benzoin (*luban jawi*)
berberine (*barbaris*)
berdache (*bardaj*)
berseem (*birsim*)
Betelgeuse (*bayt al-jauza*)
bezoar (*bazahr*)
borage (*abu arak*)
borax (*buraq*)
boron (*buraq*)
cafard (*kafir*)

calfata (*qalfatta*)
caliber (*qalib*)
caliper (*qalib*)
camphor (*kafur*)
candy (*qandi*)
Caph (*kaff al-khadib*)
carafe (*gharraf*)
carat (*qirat*)
caraway (*al-karawiya*)
carmine (*qirmiz*)
carob (*kharrub*)
carrack (*qaraqis*)
casbah (*qasba for qasaba*)
check (*sakk*)
chemistry (*al-kimiya*)
chiffon
cinnabar (*ziniafr*)
cipher (*sifr*)
civet (*zabad*)
cle (*iklid*)
Coptic (*Quft*)
coral (*qaral*)
cordevan (*Cordova*)
cornea (*curnia*)
cosine (*jayb*)
cotton (*qutn*)
crocus (*kurkum*)
Cursa (*Kursi*)
damask (*Damascus*)
decipher (*sifr*)
Deneb (*dhahab*)
Dubhe (*dhahr dubb*)
dhow (*dawa*)
dinero (*dinar*)
dourine (*darina*)

drug
dura mater
elixir (*al-iksir*)
fakeer (*faqir*)
fanfare (*farfar*)
fanfaronade (*farfar*)
fellah (*fallah*)
feloque (*fuluk*)
fustian
fomalhaut (*fum' l hut*)
gabelle (*kabala*)
gala (*khila*)
garble (*gharbal, ghirbal*)
gauze (*kazz*)
gazelle (*ghazal*)
ghoul (*ghul, from ghala*)
giraffe (*zirafah*)
harem (*harim*)
jar (*jarrah*)
jargon (*zarqun*)
jasmine (*yasmin*)
jennet (*zinata*)
julep (*ghulab*)
khamsin (*khamsun*)
kismet (*qasama*)
kohl (*kuhl*)
Kufic (*al-Kufa*)
lapis lazuli (*lazaward*)
latten (*latun*)
lemon (*laymun*)
lilac (*lilak*)
logarithm (*al-khwarizmi*)
luffa (*luf*)
lute (*al-' ud*)
macabre (*maqbarah*)

mafia (*mahya*)
magazine (*makhazin*)
Magrez (*magrizadh*)
marcasite (*marqashiti*)
mask (*maskharah*)
mattress (*matrah*)
menniges (*minningis*)
mesentary (*masarlike*)
Mizar (*mi' zar*)
mocha (*al-mukha*)
mohair (*mukhayyar*)
monsoon (*mawsim*)
mummy (*mum*)
muslin (*Mosul*)
myrrh (*murr*)
nadir (*nazir as-sam't*)
naker (*nakara*)
nucha (*nukha*)
ole (*Allah*)
orange (*naranj*)
pia mater (*umm raqiqah*)
racket (*raha*)
realgar (*rahj al-ghar*)
rebec (*rabab*)
retina
Rigel (*rijl jauzah al yusra*)
risk (*rizk*)
rook (*rukh*)
root (for square root)
safflower (*asfar*)
saffron (*za'fran*)
sal-ammoniac
saphenous (*safin*)
Saracen (*sharqiyin*)
Saros (*Sahur*)

sash (*shash*)
satin (*zaituni*)
senna (*sana*)
simar (*zamarra*)
sine (*jayb*)
sinus (*jayb*)
Sirius (*Shi'ra yamaiyya*)
sirocco (*sharaqa*)
sloop
soda (*suwayd*, from *aswad*)
sofa (*suffa*)
sorbet (*sharba*)
Spica (*simak a'zal*)
spinach (*isfanakh*)
sugar (*sukkar*)
sumac (*sumaq*)
syrup (*sharab*)
tabby (*'Attabiyyah*)
tabour (*tumboor*)
talc (*talq*)
talisman (*tilism*)
taluk (*ta'lluq*)
tamarind (*tamr hind*)
tambour (*tumboor*)
tariff (*ta'rif*)
tass (*tasa*)
tambourine (*tumboor*)
tangent and cotangent
tangerine (*Tangier*)
taraxacum (*tarashaqun*)
tare (*tarhah*)
Tauri (*Karn al-Thaur*)
tennis (*tinnis*)
theodolite (*al-'idaadah*)
traffic (*trafiq*)

turmeric (*kurkum*)
typhoon (*turan*)
Vega (*waqi*)
vizier (*wazir*)
Zaurek (*tali masaf an-nahr*)
zenith (*sumut*)
zero (*sifr*)
zircon (*zarqun*)

Table Three

Medicines Invented or Perfected by Islamic Scholars

acetic acid

acids (for cauterization)

aconite

alcohol (as an antiseptic)

aloe

ambergris

aromatic waters of herbs

aromatic waters of spices

aromatic waters of flowers

artemesia

benzoin

borage

camphor

caraway

cassia

cherry pit extract

cherry stem extract

cloves

colocynth

copper sulfate (astringent)

cubebs

cucumber seed extract

cumin

darnel (for anesthesia)

fig extract

ginger

hashish (for anesthesia)

hemlock (for anesthesia)

honey-based syrups

iron tablets (for anemia)

jalap

jasmine extract

lavender oil

lavender water

lemon peel extract

lettuce seed

mandrake

mercurial antiseptics

musk oils

myrrh

nutmeg

opium (for anesthesia)

orange blossom water

orange peel extract

oregano

pomegranate extract

rhubarb extract (laxative)

rose oil

rose water

syrup of carob (laxative)

thyme

violet water

Table Four

Country of Origin of Famous Islamic Scientists

Ibn Sina	Bukhara
ar-Razi	Persia
al-Battani	Mesopotamia
Ibn Battutah	Tangiers
Ibn Rushd	Spain
Ibn Zuhr	Spain
az-Zahrawi	Spain
Abi Usaybia	Syria
al-Jabr	Persia
Abu Mansur	Persia
al-Khwarizmi	Persia
'Umar Khayyam	Persia
Abu'l Hasan	Persia
al-Kindi	Iraq
Ibn Khaldun	Tunis
al-Ghafiqi	Spain
al-Mahusin	Syria
al-Kashi	Samarkand
Ja'far as-Sadiq	Persia
al-Uqlidisi	Syria
Ibn Juljul	Spain
al-Farabi	Persia
Ibn Butlan	Iraq
Ibn Tufayl	Spain
al-Bitruji	Spain
Ibn Bajah	Spain
Thabit bin Qurrah	Mesopotamia
al-'Urdi	Iraq
Ulug Beg	Samarkand
al-Biruni	Persia

at-Tabari ....Syria
Ibn Yunus ....Egypt
Ibn an-Nafis ....Egypt
Ibn al-Quff ....Egypt
al-Idrisi...Morocco
al-Masudi ....Iraq
al-Majusi...Persia
al-Farasi ....Persia
al-Haytham. ....Persia
Banu Musa ....Iraq
Hunayn..Iraq
Ibn al-Awan..Spain
al-Makkar ....Spain
Rashid ad-Din..Persia
Ibn Firnas ....Spain
Abu'l Wafa' ....Persia
Yaqut...Greece
al-Khazini...Greece
al-Jahiz. ....Africa

It is noteworthy that with the exception of al-Kindi not a single indigenous Arab is found on this list. The majority of the scholars were from non-Arabic countries, such as Persia, Spain, Iraq, and North Africa, illustrating the global influence of Islam. By conquering the dilapidated civilizations of the world and placing humanity under one valued philosophy, human civilization flourished. For the first time in history everyone, regardless of race or culture, had the opportunity to excel in the sciences. Every known civilization was represented, dispatching scholars to study, as well as publish, under the auspices of Islam. Other representative cultures include Turks, Greeks, Sudanese, Chinese, East Indians, Russians, and Europeans.

Table Five

List of Substances and Devices Introduced to the West by the Muslims

Foods and Herbs

artichokes
asparagus
bananas
buckwheat
cherries
cinnamon
coriander
cumin
fruit juice
ginseng
ginger
grapefruit
ice cream
mandrake extract
melons
myrrh
orange blossom water
oranges
peaches
pepper
rice
rose oil
rose water
salt
sandalwood
scammony

senna
sherbet
spinach
strawberries
sugarcane
sumac
tamarind
tangerines
turmeric
wormwood

Clothing and Textiles

satin
silk
patterned silk
striped silk
cotton underwear
three piece suits
curtains
oriental rugs
velvet
"damask" linen
fine furs
fur lined coats and boots
leather jackets
leather shoes
spinning wheels
handkerchief

table cloths
table napkins

Architectural
spine and belfry
pointed arches
windmills
ribbed vaults
ceramic tiles
mosaic tiles
decorative water fountains
stained/leaded glass

Business and Industrial
pendulums
ceramic tiles
irrigation devices
bookbinding
metal inlay
calligraphy
lateen sail
rudders
compass (for navigation)
compass (for math)
protracter
secant
sexton
quadrant
astrolabe
slide rules
rulers
sulfuric acid
nitric acid

bellows
colored glass
glass making devices
street lamps
cement
lathes
mechanical coin presses
paper money
letters of credit
flasks
postage stamps
caliper
crystal
kerosene
distillation devices
block printing (for textiles)
moveable type
printing devices
paper
reams of paper
colored paper
colored inks
steel and hemp cable
irrigation devices
cultivated flowers
self-trimming lamps
saddles
glass mirrors
solder
chronometers
weight-driven clocks
pendulum clocks
maps and almanacs

globes
planetariums
test tubes
lenses
window panes

trumpets
the music scale
pianos
organs
castanets

Medical Devices

surgical

    scissors

    forceps

    scalpels

    suture

hypodermic needles
urinary catheter
tongue depressor
cotton balls and padding
splints
casts
muslin
gauze
filter paper
plaster of Paris
topical antiseptics
artificial teeth
enema bags
seton

Musical Items

sheet music
tambourines
kettledrums
flutes
guitars

Military

firearms
wheeled cannons
crossbow
coats of mail
catapult
gunpowder
grenades

Personal, Athletic, and Entertainment

rackets
tennis balls
croquet
polo
horse races
moralistic novels
fireworks
stringed instruments
traveling bands

Table Six

Sciences Originated by the Muslims

Algebra	Mineralogy
Anesthesia	Modern Surgery
Biology	Modern Medicine
Botany	Modern Arithmetic
Cardiology	Obstetrics
Cartography	Ophthalmology
Chemistry	Optics
Dermatology	Ornithology
Ecology	Orthopedics
Embryology	Parasitology
Emergency Medicine	Pathology
Ethnography	Pediatrics
Gastroenterology	Pharmacology
Geography	Physiology
Geology	Preventive Medicine
Gynecology	Psychiatry
Horticulture	Psychology
Human Physiology	Public Health
Hydrostatics	Pulmonary Medicine
Immunology	Sociology
Internal Medicine	Toxicology
Mechanics	Trigonometry
Medical Ethics	Urology
Metallurgy	Veterinary Medicine
Meteorology	Zoology

Table Seven

Sciences Advanced by the Muslims

Acoustics	Geometry
Agronomy	Geophysics
Anatomy	Meteorology
Astronomy	Physics
Calculus	Taxonomy
Electrochemistry	Thermodynamics
Engineering	Zoology
Genetics	

Table Eight

Famous Europeans Directly Influenced by Islamic Scientists (or their writings)

Antoine Lavoisier	Emanuel Kante
Robert Boyle	Cervantes
Johannes Kepler	Petrarch
Francis Bacon	Boccacio
Tycho Brahe	Toscanelli
St. Thomas Aquinas	Archbishop of Canterbury
Dante	Biagioda Parma
Shakespeare	Filippo Brunelleschi
Vasco da Gama	Alberti
Christopher Columbus	Regiomontanus
Leonardo da Vinci	Theodoric of Freiberg
Leonardo Fibonacci	Robert Grosseteste
Roger Bacon	Adelard of Bath
Pope Syllvester II (Gerbert)	Michael Scot
Galileo	Robert of Chester
Libavius (Andreas Libau)	Alexander Neckham
Copernicus	Duns Scotus
John Napier	King Frederick II
Rene Descartes	Charlemagne
Isaac Newton	Shakespear
King Roger II (of Sicily)	Thomas Bradwardine
Berthold Swartz	Whitelo
Albertus Magnus	William Defoe
Bruno	William Harvey
Paracelsus	Bartholemeul Diaz
John Napier	Joseph Labrosse
Leonhard Rauwolf	

Table Nine

Latin Names of Famous Islamic Era Scholars

The pervasive influence of Islam upon modern civilization is easily realized by the following list. These scholars, Muslims, Christians and Jews of the Islamic Empire, were deemed by their European students as the preferred masters, superior to the ancient Greeks.

Latin Name	Arabic Name
Averroes/Aventuez	Ibn Rushd
Avenzoar	Ibn Zuhr
Avincenna	Ibn Sina
Alpetragius	al-Bitruji
Alfarabi	al-Farabi
Alhazen	al-Haytham
Alkindi	al-Kindi
Avempace	Ibn Bajjah
Albumasar	Abu Ma'shar
Albucasis	az-Zahrawi
Algorismus	al-Khwarizmi
Albategenius	al-Battani
Haly Abbas	al-Majusi
Rhases	ar-Razi
Jabir	Jabr bin Hayyan
Omar Khayyam	'Umar al-Khayyami
Azophi	as-Sufi
Mesue	al-Masawayh
Anaritius	an-Nayrizi
Messahalla	Ma sha'Allah
Haly Heben	'Ali ibn Ridwan
Haly Filius Abenragel	'Ali ibn Abi'l Rijal
Albenguefit	al-Lahmi
Maimonides	Ibn Maymun
Serapion	Ibn Sarabiyun

Table Ten

Proper Names of Famous Islamic Scholars

'Abdur-Rahman as-Sufi
'Abdul-Latif al-Baghdadi
Abu Hanifah ad-Dinawari
Abu al-Hasan bin Ibrahim al-Uqlidusi
Abu Ja'far al-Khazin
'Ali bin 'Abbas al-Majusi
Jabir bin Hayyan al-Azdi
Khalifah bin Abu al-Mahusin
'Umar al-Khayyami
Muhammad bin 'Isa al-Mahani
Nasr ad-Din at-Tusi
al-'Abbas bin Sa'id al-Jawhari
Muhyi ad-Din al-Maghribi
Ahmad bin Muhammad bin Kathir al-Farghani
Abu'Ali al-Hasan bin al-Hasan bin al-Haytham
Thabit bin Qurrah as-Sabi al-Harrani
al-Mukhtar bin 'Abdun bin Butlan
Abu Bakr Muhammad bin Zakariya ar-Razi
'Ali bin Sahl Rabban at-Tabari
Muhammad bin Qasim bin Aslam al-Ghafiqi
Abu Bakr al-Baytar
Abu Zayd Hunayn bin Ishaq al-Ibadi
Abu 'Ali al-Hunayn bin Ishaq al-Ibadi
Abu al-Qasim Khalaf bin 'Abbas az-Zahrawi
Badi' az-Zaman Isma'il bn ar-Razzaz al-Jazari
'Abd ar-Rahman bin Muhammad ibn Khaldun
Abu al-Walid Muhammad ibn Rushd
Abu Yusuf Ya'qub al-Kindi
Muhammad ibn 'Abdullah ibn Battutah
Muhammad ibn Musa al-Khwarizmi
Abu'l Fath al-Khazini

Abu 'Abd Allah Muhammad ibn Mu'adh al-Jayyani
Abu Sahl al-Kuhi
Abu Abdullah al-Battani
Abu'l Hasan 'Ali ibn abi Rijal al-Maghrabi
Qusta ibn Luqa al-Balabakki
Ibn al-Wafid al-Lahmi
Jamshid al-Kashani
Shaykh Baha ad-Din 'Amili
Mulla Muhammad Baqi Yazdi
Abu'l 'Abbas ibn Banna al-Marrakushi
Kamal ad-Din al Damiri
Ali ibn Rabban at-Tabari
Abu Sa'id as Sijzi
Abu al-Rayhan Muhammad ibn Ahmad al-Biruni
Muhammad Abu Nasr al-Farabi
Ibrahim al-Sahdi
Abu Ishaq al-Bitruji
Abu Abdullah Muhammad al-Idrisi
Abu Abdullah Yaqut
Ahmad bin at-Tayyib as-Surakhsi
Qutb ad-Din ash-Shirazi
Abu Zayd Ahmad bin Sahl al-Balkhi
Ibn al-Faqih al-Hamidhani
Ghiyath ad-Din as-Shirazi
'Abbas ibn Firnas
Abu 'Abdullah al-Mahani
Abu Ma'shar al-Balkhi
'Ali ibn Ridwan
'Ala ad-Din ibn Nafis
Afdal ad-Din al-Kashani
Abu'l Wafa' al-Buzjani
Abu Marwan ibn Zuhr
Abu'l Hasan 'Ali al-Mas'udi
Ibn Yunus al-Masri

Table Eleven

Pronunciation Guide For English-Speaking Individuals: How to Pronounce the Names of Famous Islamic Scholars

This chart is a special guide for individuals who cannot speak Arabic. It is an aid for correctly pronouncing the names of famous Islamic scholars. In Arabic there are two types of vowel sounds: short and long. Unless identified there is no way for an English speaking person to know which vowel sounds are elongated. Thus, these long sounds are represented in the following chart as a doubling of the vowel. The quotation mark (') is yet another sign to aid pronunciation. It signifies a harsh guttoral sound when pronouncing a vowel, a noise similar to a deep clearing of the throat.

Name	Pronunciation
al-Baghdadi	al-Baghdaadi
Ibn Bajjah	Ibn Baajjah
al-Battani	al-Battaanee
Ibn al-Baytar	Ibn al-Baytaar
al-Bitruji	al-Bitroojee
al-Biruni	al-Beeroonee
Ibn-Butlan	Ibn Butlaan
ad-Dinarawi	ad-Deenawaree
al-Farabi	al-Faaraabee
al-Farghani	al-Faraghaanee
Ibn Firnas	Ibn Firnaas
al-Haytham	al-Hai-tham
al-Idrisi	al-Idreesee
'Ali ibn 'Isa	'Ali ibn 'Eesaa

Hunayn ibn Ishaq	Hunayn ibn Ishaaq
al-Jahiz	al-Jaahiz
Kamal ad-Din	Kamal ad-Deen
al-Kashani	al-Kaashaanee
al-Kashi	al-Kaashee
Ibn Khaldun	Ibn Khaldoon
'Umar Khayyam	'Umar Khayyaam
al-Khazin	al-Khaazin
al-Khwarizmi	al-Khwaarizmi
al-Kindi	al-Kindee
al-Mahani	al-Mahaanee
al-Majusi	al-Majoosee
al-Mas'udi	al-Mas'oodee
Ibn Maymun	Ibn Maymoon
Ibn an-Nafis	Ibn an-Nafees
ar-Razi	ar-Raazee
Ibn Sina	Ibn-Seenaa
Ibn Tufayl	Ibn-Tufayl
at-Tusi	at-Toosee
al-Uqlidusi	al-Uqleedusee
Abu'l Wafa'	Abu'l Wafaa'
Yaqut	Yaqoot
Ibn Yunus	Ibn Yoonus
az-Zahrawi	az-Zahraawee
az-Zarqali	az-Zarqaalee

Table Thirteen

Special Contributions of Islamic Scholars to the Sciences through Mathematics

Rectification of latitudes and longitudes
Production of the first compass with movable arm
Solution of
    tri-linear figures
    cubic equations
    equations to the fourth degree
    quadratic equations
Development of the concept of square roots
Introduction of
    irrational numbers
    binomial theorem
    Arabic numerals
    sine and cosine
Invention of the Western symbol for zero
Invention of the trigonometric function *tangent*
Calculation of the radius of the earth
Use of negative numbers and symbols
Use of symbols for multiplication and addition
Application of algebra to geometry
Development of algebra into systematic science
Calculation of the third lunar variation
Calculation of the apogee of the moon and sun
Systematic use of trigonometry for astronomical calculations
Displacement of Europe's archaic Roman numerals
Precise mathematical determination of the stars' positions
Precise determination of duration of calendar year
Prediction of lunar eclipses
Creation of new theorums for integral calculus

Replacement of fractions with decimals
Production of logarithms and logarithmic tables
Determination of the obliquity of the ecliptic
Production of the first tables of versed sines and arc sines
Production of the first tables of arc contangents
Creation of trigonometry as a separate science from astronomy
Determination of the precession of the equinoxes
Determination of solar parallax
Invention and synthesis of the "sinus" theory
Calculation of distance between moon and earth
Measurement of the earth's circumference
Determination of Pascal's (i.e. Khayyam's) Triangle
Calculation of tables of sine, cosine, tangent, and contangent
Calculation of pi squared
Creation of the modern number theory

Table Fourteen

Original Contributions of Islamic Scholars in Medicine and Pharmacology

Invention of the seton
Invention of plaster of Paris as casting material
Determination of communicable nature of
    cholera
    tuberculosis
    Bubonic Plague
    smallpox
    measles
    diarrheal diseases
    gonorrhea
First accurate description of
    intestinal tuberculosis
    mediastinal abscesses
    pericarditis
    pleurisy
    esophageal cancer
    pharyngeal paralysis
    otitis media (middle ear infection)
    hydrocephalus
    hemophilia
    bowel cancer
    genetic deformities of mouth and dental arches
    smallpox
    measles
    rabies
Construction and operation of first
    emergency rooms
    trauma centers
    modern hospitals

Creation of first pharmacies
Production of first pharmacopoeias
Invention of phlebotomy for stopping cerebral hemorrhage
Treatment of venomous bites with garlic extract
Use of gastric lavage for treatment of poisoning
Invention of intra-abdominal tubes (trochars) for drainage
    of abdominal abscesses
First use of topical antiseptics
Synthesis of the drugs
Introduction of inhalation anesthesia
First treatment of psychoses with narcotics
Establishment of first Board of Medical Examiners
First systematic use of human dissection for study of
    medical anatomy
First correct description of anatomy and physiology of retina
Distinguishing of leprosy as solely a physical disease
Use of water baths for the treatment of severe fevers
Introduction of animal experimentation for determining
    drug actions and/or toxicities
First use of psychotherapy
First use of cotton for
    wound dressings
    padding casts
Introduction of silk and wool sutures for wound closure
Perfection of suture made from animal intestines (Catgut)
Invention of urinary catheters
Invention of tongue depressor
Establishment of
    first modern-style hospitals
    morning and afternoon hospital rounds
    medical schools

world's first school of pharmacy
  insane asylums
  leprosy wards
  isolation wards
  rural medical care units
Systematic use of Cesarian section
Introduction of the lithotomy position for obstetrical delivery
Creation of the science of midwifery
Introduction of squatting posture for obstetrical delivery
Creation of lithotrity, a surgical procedure for crushing
    bladder stones for removal
Recognition of scabies as being caused by mites
Treatment of scabies with organic sulfur
First treatise on surgery for trauma patients
First use of cautery for wound coagulation
Introduction of cautery as treatment for boils and skin tumors
First description of the number and specific function
    of extraocular muscles
First systematic use of cardiac drugs
First correct description of
    the pulmonary circulation
    function of heart valves
    histology of blood vessels
    anatomy of veins
    anatomy of iris and cornea
    anatomy of optic nerve
Invention of hypodermic needle
Use of hypodermic needle for suction removal of cataracts
First description of presence of sugar in blood
First observation of presence of sugar in urine of diabetics
Invention of specialized surgical knives, reactors and saws
Origination of the use of glandular extracts in the treatment

of endocrine diseases
Introduction of vaccination as a treatment for smallpox
First observation that excess consumption of dietary
    carbohydrates cause obesity
Introduction of the following surgeries
    lancing of the ear drum
    tonsillectomy
    release of urinary stricture
    lithotrity
    surgical removal of bladder stones
    cranioclasty
    paracentesis
    cataract extraction
    hernia repair
Use of ice and cold water to impede or stop hemorrhage
Invention of tourniquet
Description of paralysis as due to brain stem or
    spinal cord injury
First detailed description of anatomy of liver, spleen, kidneys,
    pancreas, intestines, and stomach
First correct account of nerve supply to viscera
Discovery of the pia mater (the thin membrane which enshrouds
the brain and spinal cord)

Table Fifteen

Devices/Items Adopted by Europeans from Islamic Civilization

The items included on this list are innovations and/or developments brought to the West exclusively via Islamic civilization. In other words, Europe received them from Islam, not from other cultures such as the Chinese, Byzantines, or Hindus. The listing, in the words of Thomas Goldstein, represents a mere portion of the "Gift of Islam."

colored paper	toothbrush
colored ink	firearms
regular ink	cannons
rolls of paper	tongue depressor
stationary	surgical instruments
bookbinding	enema bag
embossed leather	surgical suture
block printing	hypodermic needle
moveable type	syringes
embroidery	tourniquet
cotton clothing	test tubes
satin sheets	beakers
pillows	bellows
sofa	planetariums
handkerchiefs	balance scales
table napkins	tennis rackets
cotton balls	tennis balls
trousers	leather saddles
cotton padding	leather jackets
gauze	toilet waters
perfume and perfume bottles	incense

Appendix B

Setting the Record Straight

The majority of individuals believe that the sciences are exclusively the products of Western minds. It is an issue that has never before been questioned systematically. A review of any of the standard texts or encyclopedias regarding the history of science supports this statement. As these books are perused it becomes evident that the only contributors given significant mention are Europeans and/or Americans. This is particularly true of the basic texts used to teach the history of science to children, teenagers, and college students. It is hardly necessary to repeat the oft-mentioned names: Galileo, Copernicus, Brahe, Kepler, Bacon, Newton, da Vinci, Harvey, Boyle, Franklin, Einstein, etc. The unavoidable conclusion is that modern science and technology is European/American in origin and that major contributions to the development of modern science by other cultures is minimal, perhaps non-existent. Most texts give little or no mention of the advancements made by ancient Indian, Persian, Chinese, or, particularly, Muslim scholars.

Western civilization has made invaluable contributions to the development of the sciences. However, so have numerous other cultures. The problem is that Westerners have long been credited with discoveries made centuries before by Islamic scholars. Thus, for many innovations, including monumental ones, the actual discover or inventor has been omitted from the historical record, and, instead, credit has been fraudulently given to others. What's more, this has occurred in thousands of instances. The gravity of these errors is emphasized by the fact that the majority of the basic sciences were invented by non-Europeans. Chemistry, physics, optics, algebra, trigonometry, basic arithmetic, mathematical astronomy, modern medicine, pharmacology, geography, and geology are all non-Western inventions. Yet, few if any individuals, even history teachers, have any clue of this fact. The fact that the West owes the creation of the sciences to other civilizations is documented by the world's top historians of science. George Sarton, Harvard's greatest scientific historian, states that modern Western medicine did not originate from Europe and that it arose instead strictly from the (Islamic) Orient. Thomas Goldstein, one of America's premier Medieval historians, says that all of the detail oriented sciences are strictly products of Islam.

The data in this section concerning dates, names, and topics of Western advances has been derived primarily from four main sources: *World Book Encyclopedia, Encyclopedia Britannica, The Popular Science Encyclopedia of Science*, and Isaac Asimov's 700 page book, *Chronology of Science and Discovery*. Supportive data for the accomplishments of Islamic scholars is derived from the miscellaneous references listed in the bibliography of this book.

What is Taught: The Greeks were the developers of trigonometry.

What Should be Taught: Trigonometry remained largely a theoretical science among the Greeks. It was developed to a level of modern perfection by Muslim scholars, although the weight of the credit must be given to al-Battani and at-Tusi. The words describing the basic functions of this science, sine, cosine, and tangent, are all derived from Arabic terms. Thus, original contributions by the Greeks in trigonometry were minimal. Furthermore, according to Durant both European and Chinese trigonometry are of Arabic origin. DeVaux states that both plane and spherical trigonometry were "indisputably" discovered by the Muslims.

What is Taught: The compass was invented by the Chinese, who may have been the first to use it for navigational purposes sometime between 1000 and 1100 A.D. The earliest reference to its use in navigation was by the Englishman, Alexander Neckam (1157-1217).

What Should be Taught: Muslim geographers and navigators learned of the magnetic needle, possibly from the Chinese, and were the first to use magnetic needles in navigation. They invented the compass and taught Westerners, as well as Orientals, how to use it. European navigators relied upon Muslim pilots and their instruments when exploring unknown territories. Gustav Le Bon claims that the magnetic needle and compass were entirely invented by the Muslims and that the Chinese had little to do with it. Neckam, as well as the Chinese, probably learned of it from Muslim traders. It is noteworthy that the Chinese improved their navigational expertise after they began interacting with the Muslims during the 8th century. The fact is the first Chinese description describes the *Arabic* use of this device.

What is Taught: The use of rudders for ships was developed in the Middle Ages by European ship builders. Prior to their invention ships were steered by an individual, who held a broad oar into the water at the back of the boat.

What Should be Taught: Muslim ship builders developed rudders sometime between the 11th and 12th centuries. It is likely that the Crusaders brought this technology to the West. Europeans began building rudders about 1240 A.D.

What is Taught: The 13th century English scholar Roger Bacon first mentioned glass lenses for improving vision. At nearly the same time eyeglasses could be found in use both in China and Europe.

What Should be Taught: During the 9th century Ibn Firnas of Islamic Spain invented eyeglasses, and they were manufactured and sold throughout Spain for over two centuries. Europeans, as well as the Chinese, learned about them from Muslim traders. Any mention of eyeglasses by Roger Bacon was simply a regurgitation of the work of al-Haytham (d. 1039), whose research Bacon frequently referred to. There is absolutely no doubt that glass lenses were invented by Islamic scientists.

What is Taught: During the 17th century the German astronomer Johannes Kepler revolutionized astronomy by determining that the sun is the center of the solar system and, more importantly, that the planetary orbits are elliptical instead of circular.

What Should be Taught: According to the *Dictionary of Scientific Biography* Islamic Spain's az-Zarqali categorically stated that the orbit of mercury is elliptical instead of circular some 500 years before Kepler. Hundreds of Islamic astronomers documented that the sun is the center of the solar system, and, as Welty states, this was common knowledge among Muslim astronomers. It is well known that Kepler, as well as Copernicus, was influenced by az-Zarqali.

What is Taught: Gunpowder was developed in the Western world as a result of Roger Bacon's work in 1242. The first usage of gunpowder in weapons occurred when the Chinese fired it from bamboo shoots in attempt to frighten Mongol conquerors. They produced it by adding sulfur and charcoal to saltpeter.

What Should be Taught: The Chinese developed saltpeter for use in fireworks, but they knew of no tactical military use for gunpowder, nor did they invent its formula. Research by Reinuad and Favé has clearly shown that gunpowder was formulated initially by Muslim chemists. Furthermore, these historians claim that the Muslims developed the first fire-arms. Notably, Muslim armies used grenades and other weapons in their defense of Algericus against the Franks during the 14th century. Jean Mathes indicates that Muslim rulers had stock-piles of grenades, rifles, crude cannons, incendiary devices, sulfur bombs, and pistols decades before such devices were used in Europe and provided the pictures to prove it. It is notable that the term arsenal is of Arabic origin (from *darcina' ah*). The first mention of a cannon was in an Arabic text around 1300 A.D. Roger Bacon learned of the formula for gunpowder from Latin translations of Arabic books. He produced nothing original in this regard.

What is Taught: No improvement was made in the astronomy of the ancients during the Middle Ages regarding the motion of planets until the 13th century. Then, Alphonso the Wise of Castile (Middle Spain) invented the *Aphonsine Tables*, which were more accurate than Ptolemy's.

What Should be Taught: As early as the 9th century Muslim astronomers made hundreds of improvements upon Ptolemy's findings. They were the first astronomers to dispute his archaic ideas. In their critique of the Greeks they synthesized proof that the sun is the center of the solar system and that the orbits of the earth and other planets might be elliptical. They produced hundreds of highly accurate astronomical tables and star charts. Many of their calculations are so precise that they are regarded as contemporary. The Alphonsine Tables are little more than copies of works on Islamic astronomy transmitted to Europe via Spain, i.e. the *Toledo Tables*. It is interesting to note that Alphonso was given his surname, i.e. The Wise, because he knew Arabic and studied Islamic books.

What is Taught: The science of geography was revived during the 15th-17th centuries, when the ancient works of Ptolemy were discovered. The Crusades and the Portuguese/Spanish expeditions also contributed to this reawakening. During this period the first scientifically based monographs on geography were produced by Europe's scholars.

What Should be Taught: Muslim geographers produced untold volumes of books on the geography of Europe, Africa, Asia, India, China, and the East Indies during the 8th through 15th centuries. These writings included the world's first geographical encyclopedias, almanacs, and road maps. The masterpieces of the Islamic geographers provide a detailed view of the geography of the ancient world. The works were organized and systematic. The Muslim geographers of the 9th through 15th centuries far exceeded the output by Europeans regarding the geography of these regions well into the 19th century. No one, not even modern historians, has ever matched their descriptions. The Crusades led to the destruction of educational institutions, their scholars, and books. They brought nothing substantive regarding geography to the Western world.

What is Taught: The first mention of man in flight was by Roger Bacon, who drew a flying apparatus. Leonardo da Vinci also conceived of airborne transport and drew several prototypes.

What Should be Taught: Ibn Firnas of Islamic Spain invented, constructed, and tested a flying machine in the 800s A.D. Roger Bacon learned of flying machines from Arabic references to Ibn Firnas' machine. The latter's invention antedates the drawings of Bacon by 700 years and da Vinci by some 500 years.

What is Taught: During the 16th century Nicholas Copernicus, a Polish monk, revolutionized astronomy by publishing the heliocentric theory. He was the first to argue that the sun is the center of the solar system and that the earth revolves around it.

What Should be Taught: In the 14th century at-Tusi and colleagues, working and publishing at the massive, elegant Maragha Observatory in what is now known as Iraq, totally renovated ancient astronomy into a modern science. As Welty makes clear the heliocentric theory was routine knowledge among Islamic astronomers. Copernicus' theories were elementary among Muslim astronomers. He added nothing substantive to their findings.

What is Taught: During the 17th century the Englishman William Harvey revolutionized the science of physiology, being the first individual to establish the circulation of blood.

What Should be Taught: As early as the 10th century A.D. it was standard knowledge among Islamic surgeons that blood circulates. Ibn Zuhr (10th century) thoroughly and correctly described the function of the heart as a one way pump, the first such description in history. During the 13th century Ibn an-Nafis thoroughly described the flow of blood from the heart into the lungs, i.e. the pulmonary circulation. He and Ibn al-Quff (14th century) documented blood flow from the heart all the way through the capillaries. Harvey was an admitted student of Islamic books. While some of his findings were original, it was the Muslims, not William Harvey, who discovered circulation, 300 years before Harvey was born.

What is Taught: Glass mirrors were first produced in 1291 in Venice.

What Should be Taught: Glass mirrors were in use in Islamic Spain as early as the 11th century. The Venetians learned of the art of fine glass production from Syrian artisans during the 9th and 10th centuries.

What is Taught: Purified alcohol, made through distillation, was first produced by Arnau de Villanova, a Spanish alchemist, in 1300 A.D.

What Should be Taught: Numerous Muslim chemists produced medicinal-grade alcohol through distillation as early as the 10th century. Furthermore, they were the first to manufacture on a large scale distillation devices for use in chemistry and industry. They used alcohol as a solvent and antiseptic. Today, it is used for these purposes.

What is Taught: Europeans landed on the Canary Islands, a group of small islands off the shores of Western Africa, in 1312 A.D.

What Should be Taught: Muslim explorers discovered these islands in 999 A.D.

What is Taught: Until the 14th century the only type of clock available was the water clock. In 1335 a large mechanical clock was erected in Milan, Italy. This was possibly the first weight driven clock.

What Should be Taught: In the 10th through 12th centuries a variety of mechanical clocks, both large and small, were produced by Spanish Muslim engineers. The knowledge of their production was transmitted to Europe through Latin translations of Islamic books on mechanics. These clocks were weight-driven. Designs and illustrations of epicyclic and segmental gears were provided. One such clock included a mercury escapement. The latter type was directly copied by Europeans during the 15th century. In addition, according to Durant during the 9th century Ibn Firnas of Islamic Spain invented a watch-like device, which kept accurate time. The Muslims also constructed a variety of highly accurate astronomical clocks for use in their observatories. When Europeans conquered Islamic Toledo (in Spain) during the 11th century they were ignorant of the value of time. However, in Toledo they found working clocks. They disassembled them in order to determine how they worked. However, after reassembling them they failed to work.

What is Taught: The concept of quarantine was first developed in 1403. In Venice a law was passed preventing strangers from entering the city until a certain waiting period had elapsed. If, by then, no sign of illness could be found, they were allowed in.

What Should be Taught: The concept of quarantine was first introduced in the 7th century A.D. by the Prophet Muhammad, who wisely warned against entering or leaving a region suffering from plague. As early as the 10th century Muslim physicians innovated the use of isolation wards for individuals suffering with communicable diseases.

What is Taught: The magnetic compass and the astrolabe made the historic voyage of Christopher Columbus in 1492 possible.

What Should be Taught: While the above is true it must be emphasized that it was the Muslims alone who developed these devices for navigational purposes. However, their components were improved upon by European navigators.

What is Taught: Isaac Newton invented the First Law of Gravity during the 17th century after an accident of fate: an apple fell on his head while he was sitting under an apple tree. This lead him to conclude that all objects are attracted to the earth's center.

What Should be Taught: The First Law of Gravity was invented in totality by Islamic physicists, notably al-Biruni, Ibn Sina, ar-Razi, Ibn Bajjah, and Qutb ad-Din, 600 years prior to Newton. The apple story has recently been debunked as a fable.

What is Taught: Movable type and the printing press were invented in the West by Johannes Gutenberg of Germany during the 15th century.

What Should be Taught: In 1454 Gutenberg developed the most sophisticated printing press of the Middle Ages. However, Durant suggests that the idea of printing with wooden blocks first originated in Cairo during the 11th-12th centuries. Moreover, movable brass type was in use in Islamic Spain over 100 years prior to Gutenberg, and that is where the West's first printing devices were made.

What is Taught: Leonardo da Vinci (16th century) fathered the science of geology when he noted that fossils found on mountains indicated a watery origin of the earth.

What Should be Taught: Al-Biruni (11th century) made precisely this observation and added much to it, including a huge book on geology, hundreds of years before da Vinci was born. Ibn Sina noted this as well (see pages 133-135). This fact establishes both al-Biruni and Ibn Sina as not merely the fathers but, rather, the founders of geology. It is probable that da Vinci first learned of this concept from Latin translations of Islamic books. He added nothing substantial to their findings.

What is Taught: In a major feat of navigation Vasco da Gama rounded the Cape of Good Hope and reached Calicut, India, in May of 1498.

What Should be Taught: Vasco da Gama was assisted on the final phase of his trek by a Muslim navigator, who guided him through the difficult passage from the east coast of Africa to his destination in southern India. Without the help of the expert Muslim navigator, along with his precision instruments, successful navigation by da Gama would have been unlikely.

What is Taught: The difficult cubic equations (x to the third power) remained unsolved until the 16th century, when Niccolo Tartaglia, an Italian mathematician, solved them.

What Should be Taught: Cubic equations, as well as numerous equations of even higher degrees, were solved with ease by Muslim mathematicians as early as the 10th century.

What is Taught: The concept that numbers could be less than zero, that is negative numbers, was unknown until 1545, when Geronimo Cardano introduced the idea.

What Should be Taught: Muslim mathematicians introduced negative numbers for use in a variety of arithmetic functions at least 400 years prior to Cardano.

What is Taught: In 1545 the scientific use of surgery was advanced by the French surgeon Ambroise Paré. Prior to him surgeons attempted to stop bleeding through the gruesome procedure of searing wounds with boiling oil. Paré stopped the use of hot oils and began ligating arteries. He is considered the "father of rational surgery." Paré was also one of the first Europeans to condemn such grotesque "surgical" procedures as trepanning (see page 66).

What Should be Taught: Islamic Spain's illustrious surgeon az-Zahrawi (d. 1013) began ligating arteries with fine sutures over 500 years prior to Paré. He perfected the use of *Catgut*, that is suture made from animal intestines. Additionally, he instituted the use of cotton plus wax to plug bleeding wounds. The full details of his works were made available to Europeans through Latin translations. Despite this, barbers and herdsmen remained the primary individuals practicing the "art" of surgery for nearly six centuries after az-Zahrawi. Paré himself was a barber, albeit more skilled and conscientious than the average ones.

Az-Zahrawi's legacy included the publication of dozens of books. His most famous work is a 30 volume treatise on medicine and surgery. His books contain sections on preventive medicine, nutrition, cosmetics, drug therapy, surgical technique, anesthesia, pre- and post-operative care, as well as drawings of some 200 surgical devices, many of which he invented. The refined and scholarly az-Zahrawi must be regarded as the father and founder of rational surgery, not the uneducated—and unpublished—Paré.

What is Taught: In the 17th century the pendulum was developed by Galileo when he was a teenager. He noticed a chandelier swaying as it was being blown by the wind. As a result, he rushed home and invented the pendulum.

What Should be Taught: The pendulum was discovered by the Egyptian scholar Ibn Yunus al-Masri during the 10th century, who was the first to study and document its oscillatory motion. Its value for use in clocks was introduced by Muslim physicists during the 15th century. Galileo learned of it from Latin translations of Islamic texts, which were available to him at the Papal libraries.

What is Taught: The use of decimal fractions in mathematics was first developed by a Dutchman, Simon Stevin, in 1589. He helped advance the mathematical sciences by replacing the cumbersome fractions, for instance, 1/2, with decimal fractions, for example, 0.5.

What Should be Taught: Muslim mathematicians were the first to utilize decimals instead of fractions on a large scale. Al-Kashi's book, *Key to Arithmetic*, was written at the beginning of the 15th century and was the stimulus for the systematic application of decimals to whole numbers and fractions. It is highly probably that Stevin imported the idea to Europe from al-Kashi's work.

What is Taught: The first man to utilize *algebraic symbols* was the French mathematician, Francois Vieta. In 1591 he wrote an algebra book describing equations with letters such as the now familiar x's and y's. Asimov says that this discovery had an impact similar to the progression from Roman numerals to Arabic numbers.

What Should be Taught: Muslim mathematicians, the inventors of algebra, introduced the concept of using letters for unknown variables in equations as early as the 9th century A.D. Through this system they solved a variety of complex equations, including quadratic and cubic equations. They used symbols to develop and perfect the binomial theorem. Through their use of symbols they converted mathematics from complexities understood by only the elite to a simple science available to everyone.

What is Taught: The first *pharmacopoeia* (book of medicines) was published by a German scholar in 1542. According to *World Book Encyclopedia* the science of pharmacology was begun in the 1900s as an derivation of chemistry due to the analysis of crude plant materials. After isolating the active ingredients from plants, chemists realized their medicinal value.

What Should be Taught: According to the eminent scholar of Arabic history, Phillip Hitti, the Muslims, not the Greeks or Europeans, wrote the first modern pharmacopoeia. The science of pharmacology was originated by Muslim physicians during the 9th century. They developed it into a highly refined and exact science. Muslim chemists, pharmacists, and physicians produced hundreds of drugs and/or crude herbal extracts one thousand years prior to the supposed birth of pharmacology. During the 14th century Ibn Baytar wrote a monumental pharmacopoeia listing some 1400 different drugs. Hundreds of other pharmacopoeias were published during the Islamic Era. It is likely that the German work is a derivation of that by Ibn Baytar, which was widely circulated in Europe.

What is Taught: Galileo (17th century) was the world's first great experimenter.

What Should be Taught: The 11th century scholar al-Biruni was the world's first great experimenter. He wrote over 200 books, many of which discuss his precise experiments. His literary output in the sciences amounts to some 13,000 pages, far exceeding that written by Galileo or, for that matter, Galileo and Newton combined.

What is Taught: Isaac Newton's 17th century study of lenses, light, and prisms forms the foundation of the *modern science of optics*.

What Should be Taught: In the 11th century al-Haytham determined virtually everything that Newton advanced regarding optics centuries prior and is regarded by numerous authorities as the "founder of optics." There is little doubt that Newton was influenced by him.

Al-Haytham was the most quoted physicist of the Middle Ages. His works were utilized and quoted by a greater number of European scholars during the 16th and 17th centuries than those of Newton and Galileo combined. Incredibly, fully 600 years after they were written al-Haytham's books continued to exert a direct influence upon the progression of European science, since Newton and Galileo, as well as Kepler, were heavily influenced by them.

What is Taught: During the 17th century Isaac Newton made a major breakthrough in physics by discovering that white light consists of various rays of colored light.

What Should be Taught: This discovery was made in its entirety by al-Haytham (llth century) and Kamal ad-Din (14th century). The fact is al-Haytham was the first to bend light through glass, thereby discovering the various colors of the light spectrum 700 years before Newton.

What is Taught: In 1614 John Napier invented logarithms and logarithmic tables.

What Should be Taught: Muslim mathematicians invented logarithms and produced logarithmic tables several centuries prior to Napier. As early as the 13th century such tables were common in the Islamic world. The word logarithm itself originates from the name of Islam's most famous mathematician, al-Khwarizmi. His name was translated during the 12th century into algorism from which the word logarithm was later derived.

What is Taught: During the 17th century Rene Descartes made the discovery that algebra could be used to solve geometrical problems. As a result, he greatly advanced the science of geometry.

What Should be Taught: Mathematicians of the Islamic Empire accomplished precisely this as early as the 9th century A.D. Thabit bin Qurrah was the first to do so, and he was followed by Abu'l Wafa', whose 10th century book utilized algebra to advance geometry into an exact and simplified science.

What is Taught: During the early 17th century William Harvey discovered that blood circulates. He was the first to correctly describe the function of the heart, arteries, and veins. Rome's Galen had presented erroneous ideas regarding the circulatory system, and Harvey was the first to determine that blood is pumped throughout the body via the action of the heart and the venous valves. Therefore, he is regarded as the *founder of human physiology.*

What Should be Taught: In the 10th century Islam's ar-Razi wrote an in-depth treatise on the venous system, accurately describing the function of the veins and their valves. Ibn an-Nafis and Ibn al-Quff (13th century) provided complete documentation that the blood circulates and correctly described the physiology of the heart and the function of its valves 300 years before Harvey. William Harvey was a graduate of Italy's famous Padua University at a time when the majority of its curriculum was based upon Ibn Sina's and ar-Razi's textbooks. Because of his strikingly accurate treatise on the mechanism of circulation Ibn Nafis, not William Harvey, must be regarded as the founder of human physiology.

What is Taught: During the 17th century Isaac Newton developed the binomial theorem, which is crucial for the study of algebra.

What Should be Taught: Hundreds of Muslim mathematicians utilized and perfected the binomial theorem. They initiated its use for the systematic solution of algebraic problems during the 10th century (or prior).

What is Taught: In the 17th century Robert Boyle originated the science of chemistry.

What Should be Taught: A variety of Muslim chemists, including ar-Razi, al-Jabr, al-Biruni, and al-Kindi, performed scientific experiments in chemistry some 700 years prior to Boyle. Durant writes that the Muslims introduced the experimental method to this science. Humboldt, Bernal, and Dampier regard the Muslims as the founders of chemistry.

What is Taught: The first man to classify the races was the German Johann F. Blumenbach, who divided mankind into white, yellow, brown, black, and red peoples.

What Should be Taught: Muslim scholars of the 9th through 14th centuries invented the science of ethnography. A number of Muslim geographers classified the races, writing detailed explanations of their unique cultural habits and physical appearances. They wrote thousands of pages on this subject. Blumenbach's works were insignificant in comparison.

What is Taught: The scientific use of antiseptics in surgery was discovered in 1865 by the British surgeon Joseph Lister.

What Should be Taught: As early as the 10th century Muslim physicians and surgeons applied purified alcohol to wounds as an antiseptic agent. Surgeons in Islamic Spain utilized special methods for maintaining antisepsis prior to and during surgery.

They also originated specific protocols for maintaining hygiene during the post-operative period. Their success rate was so high that dignitaries throughout Europe came to Cordova, Spain, to be treated at what was comparably the "Mayo Clinic" of the Middle Ages. These Islamic surgeons, not Joseph Lister, were the inventors of surgical antiseptics.

What is Taught: The discovery of the scientific use of drugs in the treatment of specific diseases was made by Paracelsus, a Swiss-born physician, during the 16th century. He is also credited with being the first to use practical experience as a determining factor in the treatment of patients rather than relying exclusively on the works of the ancients.

What Should be Taught: Ar-Razi, Ibn Sina, al-Kindi, Ibn Rushd, az-Zahrawi, Ibn Zuhr, Ibn Baytar, Ibn al-Jazzar, Ibn Juljul, Ibn al-Quff, Ibn an-Nafis, al-Biruni, Ibn Sahl, and hundreds of other Muslim physicians mastered the science of drug therapy for the treatment of specific symptoms and diseases. The word "drug" is derived from Arabic. Their use of practical experience and careful observation was extensive.

Muslim physicians were the first to criticize ancient medical theories and practices. Ar-Razi devoted an entire book as a critique of Galen's anatomy. The works of Paracelsus are insignificant compared to the vast volumes of medical writings and original findings accomplished by the medical giants of Islam. He added nothing significant to their findings.

What is Taught: The first sound approach to the treatment of disease was made by a German, Johann Weger, in the 1500s.

What Should be Taught: Harvard's George Sarton says that modern medicine is entirely an Islamic development and that the Muslim physicians of the 9th through 12th centuries were precise, scientific, rational, and sound in their approach. According to Osler the descriptions of diseases by Islamic physicians are essentially modern. Johann Weger was among thousands of European physicians during the 15th through 17th centuries who were taught the medicine of ar-Razi and Ibn Sina, whose texts, which described hundreds of diseases in detail, were authoritative throughout Europe. He contributed nothing original to their findings.

What is Taught: The concept of the finite nature of matter was originally introduced by Antione Lavoisier during the 18th century. He discovered that although matter may change its form or shape, its mass always remains the same. Thus, for instance, if water is heated to steam, if salt is dissolved in water, or if a piece of wood is burned to ashes, the total mass remains unchanged.

What Should be Taught: The principles of this discovery were elaborated in detail centuries before Lavoisier by Islamic Persia's great scholar, al-Biruni, who died about 1050 A.D. Lavoisier was a disciple of the Muslim chemists and physicists. He quoted their books frequently.

What is Taught: Paul Ehrlich (19th century) is the originator of *drug chemotherapy*, that is the use of specific drugs to kill microbes.

What Should be Taught: Muslim physicians used a variety of specific substances to destroy microbes. They applied sulfur topically specifically to kill the scabies mite. They used antiseptics to sterilize the hands prior to surgery. Furthermore, they were the first to apply purified alcohol during surgery for preventing infections. Ar-Razi (10th century) used mercurial compounds as topical antiseptics. Clearly, the Muslims innovated specific drug chemotherapy *over 1000 years prior to Ehrlich.*

What is Taught: The first surgery performed under inhalation anesthesia was conducted by C.W. Long, an American, in 1845.

What Should be Taught: Six hundred years prior to Long Islamic Spain's az-Zahrawi and Ibn Zuhr, among other Muslim surgeons, performed hundreds of surgeries under inhalation anesthesia. They administered narcotic-soaked sponges, which were placed over the face. Thus, they were indisputably the inventors of inhalation anesthesia.

What is Taught: Modern anesthesia was invented in the 19th century by Humphrey Davy and Horace Wells.

What Should be Taught: Modern anesthesia was discovered, mastered, and perfected by Muslim anesthetists 900 years before the advent of Davy and Wells. They utilized oral as well as inhalant anesthetics. Furthermore, they developed special procedures for the management and protection of anesthetized patients during the post-operative period. Thus, the Muslims invented modern anesthesia as we know it today and were the world's first true anesthetists.

What is Taught: The Italian Giovanni Morgagni is regarded as the father of pathology, because he was the first to correctly describe the nature of disease.

What Should be Taught: Islam's surgeons were the first pathologists. They fully realized the nature of disease and described a variety of diseases in modern detail. Ibn Zuhr correctly described the nature of pleurisy, tuberculosis, and pericarditis. Az-Zahrawi accurately documented the pathology of hydrocephalus (water on the brain) and other congenital diseases. Ibn al-Quff and Ibn an-Nafis gave perfect descriptions of the diseases of circulation.

Other Muslim surgeons gave the first accurate descriptions of certain malignancies, including cancer of the stomach, bowel, and esophagus. These surgeons were the originators of pathology, not Giovanni Morgagni.

What is Taught: The first mention of the geological formation of valleys was in 1756, when Nicolas Desmarest proposed that they were formed over a long periods of time by streams.

What Should be Taught: Ibn Sina and al-Biruni made precisely this discovery during the 11th century (see pages 133-135), fully 700 years prior to Desmarest.

What is Taught: Medical treatment for the insane was modernized by Philippe Pinel, when in 1793 he operated France's first insane asylum.

What Should be Taught: As early as the 11th century Islamic hospitals maintained special wards for the insane. They treated them kindly and presumed their diseases were real at a time when the insane were routinely burned alive in Europe as witches and sorcerers. A curative approach was taken for mental illness, and for the first time in history the mentally ill were treated with supportive care, drugs, and psychotherapy. Every major Islamic city maintained an insane asylum, where patients were treated at no charge. In fact, the Islamic system for the treatment of the insane was superior to the current model, as it was more humane and was highly effective as well.

What is Taught: During the 16th century Paracelsus invented the use of opium extracts for anesthesia.

What Should be Taught: Muslim physicians introduced the anesthetic value of opium derivatives during the 9th through 10th centuries. Opium was originally used as an anesthetic agent by the Greeks. Islamic pharmacologists developed new and innovative techniques to refine and administer it, including the using it as a narcotic during surgery. Paracelus was a student of Ibn Sina's works from which it is virtually assured that he derived this idea. This is clearly indicated by the fact that Paracelsus regarded Ibn Sina his "Arabic (scientific) father."

What is Taught: Kerosene was first produced by the Englishman, Abraham Gesner, in 1853. He distilled it from asphalt.

What Should be Taught: Muslim chemists produced kerosene on a large scale as a distillate from petroleum products over 1,000 years prior to Gesner (see *Encyclopedia Britannica* under the heading, Petroleum).

What is Taught: During the 17th century Peter Chamberlen of England was the first physician to use forceps during delivery.

What Should be Taught: During the 10th century A.D. Islamic Spain's az-Zahrawi invented several types of forceps and used them during difficult deliveries with great success. These monumental accomplishments make it clear that az-Zahrawi essentially invented the science of modern obstetrics, *700 years before Chamberlen.*

What is Taught: Modern fashion, entertainment, and sophistication are entirely creations of modern Europe and America, wherein the refinements of life were invented.

What Should be Taught: Islamic Spain produced the world's first fashion designers, who invented three piece suits, cotton clothing, fine silk, shirts with collars, the monographed handkerchief, embossed leather, the table napkin, and even trousers. Durant gives examples of Medieval European rulers emulating Islamic sophistication by wearing robes embroidered with Arabic calligraphy. Chefs in Islamic Spain created the five course meal and published the world's first books on fine dining and manners. The concept of the gentleman arises from Islamic Spain. They even invented ice cream. Europe learned of these and other social refinements via the translation of Islamic works as well as through direct interactions. According Henderson during the 11th century when the illiterate Medieval Europeans first captured an Islamic city intact they were so overwhelmed by its sophistication and technology that "they could not believe what they saw."

Bibliography

1. Abercrombie, Thomas J. 1991. Ibn Battuta, Prince of the Travelers. *National Geographic*, December.

2. Ahsan, M. M. 1979. *Social Life under the Abbasids.* Longman Group: London.

3. Ali, Ameer. 1961. *The Spirit of Islam.* London: Christophers, pp. 361-390.

4. Asad, M. 1984. *The Message of the Qur'an.* Gibraltar: Dar al-Andalus.

5. Al-Akkad, A. M. (no date). *The Arab's Impact on European Civilization.* Cairo: S.O.P. Press.

6. Arnold, T. and A. Guillaume. *The Legacy of Islam.* London: Oxford University Press.

7. Asimov, I. A. 1989. *Asimov's Chronology of Science and Discovery.* Grand Rapids: Harper & Row.

8. Athar, S. (ed). *Islamic Medicine.* International Publ.

9. Balyaz, H. M. 1976. *Muhammad and the Course of Islam.* Oxford: G. Ronald, pp. 290-300.

10. Bammate, Haider. 1962. *Muslim Contribution to Civilization.* Crescent Publications.

11. Beg, M. A. J. 1983. *Arabic Loan-Words in Malay.* Kuala Lumpur: Univ. Malay Press. pp. 46-48.

12. Bernal, J. D. *Science in History.* (vol. 5). Cambridge, Mass: MIT Press.

13. Boorstin, Daniel J. 1983. *The Discoverers.* New York: Random House.

14. Braudel, J. D. 1949. *The Mediterranean and the Mediterranean World in the Age of Philip II.* New York: Harper & Row, Vol. 2.

15. Briffault, R. 1938. *The Making of Humanity.* London.

16. *Encyclopedia Britannica.* 1973. Chicago: William Bently, Publ. Vols. 1, 3, 10, 17.

17. Brook, Christopher. 1969. *Twelfth Century Renaissance.* Thames & Hudson.

18. Brown, P. 1978. *The Making of Late Antiquity.* Harvard Univ. Press.

19. Browne, E. G. 1921. *Arabian Medicine.* London.

20. Bucaille, M. 1980. *The Bible, Qur'an, and Science.* Paris.

21. Bukhari. (Dr. M. Khan, translator). *Sahih al-Bukhari* (the definitive statements of Prophet Muhammad).

22. Burk, J. 1985. *The Day the Universe Changed.* Boston: Little Brown & Co., pp. 36-44; 108-09; 195-220.

23. Butterfield, H. 1957. *The Origins of Modern Science.* New York: Macmillan Co.

24. Campbell, D. 1926. *Arabian Medicine and Its Influence on the Middle Ages.* London, (Vols. 1 & 2).

25. Campbell, J (ed). 1962. *The Arabian Knights.* Viking Press.

26. Crombie, A. A. 1959. *Augustine to Galileo.* Vol. I & II. Harmondsworth: Penguin.

27. Crombie, A. C. 1971. *Robert Grosseteste and the Origins of Experimental Science, 1100-1700.* Oxford: Claredon Press.

28. Daniel, Norman. 1979. *The Arabs and Medaevel Europe.* Longman Group LTD: Harlow, England.

29. Dapier, S. W. 1961. *A History of Science.* London: Cambridge Univ. Press.

30. De Vaux, Carra. 1921. *The Philosophers of Islam.* Paris.

31. Dunlop, D. M. 1958. *Arabic Science in the West.* Karachi: Pakistan Historical Society.

32. Dunlop, D. M. 1971. *Arabic Civilization to AD 1500*. Longman Group: Harlow, England.

33. Durant, Will. 1950. *The Age of Faith*. New York: Simon and Schuster.

34. Faber, E. 1961. *Great Chemists*. New York: Int. Publ., Inc., pp. 3-38.

35. Fowler, W. S. 1962. *The Development of the Scientific Method*. Oxford: Pergamom Press.

36. Garrison, F. H. 1927. *An Introduction to the History of Medicine*. Philadelphia: W. B. Saunders Co., pp. 126-248.

37. Gibb, H. A. R. 1949. *Modern Tendencies in Islam*. Paris.

38. Gibb, H. A. R. (ed) 1958-62. *The Travels of Ibn Battuta 1325-1354*. Cambridge Univ. Press.

39. Goldstein, B. 1971. *Al-Bitruji: On the Principles of Astronomy*. New Haven and London: Vol. 1 & 2.

40. Goldstein, Thomas. 1980. *Dawn of Modern Science*. Boston: Houghton Mifflin Co., pp. 92-129.

41. Haddad, S. I. and A. A. Khairallah. 1936. A forgotten chapter in the history of the circulation of the blood. *Annals of Surgery*. 104:1-8.

42. Haddad, S. 1942. Arabian Contribution to Medicine. *Ann. Med. History*. 3:60.

43. Hamarneh, S. K. 1964. *Bibliography on Medicine and Pharmacy in Medieval Islam*. Stuttgart.

44. Hamarneh, S. K. 1975. *Catalogue of Arabic Manuscripts on Medicine and Pharmacology at the British Library*. Cairo: Les Editions Universitaries d'Egypte (with the Smithsonian Institute).

45. Hamarneh, S. K. and M. Anees (ed). 1985. *Health Sciences in Early Islam*. San Antonio: Noor Health Foundation.

46. Hanafi, M. A.1969. *A Survey of Muslim Institutions and Culture*. Lahore, Pakistan: Sh. Muhammad Ashraf.

47. Haskins, C. H. 1923. *The Rise of Universities*. Cornell Univ. Press.

48. Haskins, Charles H. 1924. *Studies in the History of Medical Science*. Cambridge: Harvard Univ. Press.

49. Hayes, J. R. (ed). 1983. *The Genius of Arab Civilization*. Cambridge, Mass: MIT Press.

50. Haykal, M. H. 1976. *The Life of Muhammad*. North American Trust Publ.

51. Hennessee, O. M. and B. R. Cook. 1989. *ALOE: Myth, Magic and Medicine*. Universal Graphics.

52. Hill, J. W. 1988. *Chemistry for Changing Times*. New York: Macmillan Co., pp. 3.

53. Hill, D. R. 1993. *Islamic Science and Engineering*. England: Edinburgh Univ. Press.

54. Hitti, P. 1943. *The Arabs: A Short History*. Princeton: Princeton Univ. Press.

55. Hitti, P. K. 1961. *History of the Arabs*. New York: Macmillan & Co.

56. Hitti, P. K. 1968. *The Makers of Arab History*. New York: St. Martin's Press.

57. Hogben, L. 1960. *Mathematics in the Making*. Crescent Books.

58. Hourani, George. 1963. *Arab Seafaring*. Verry Lawrence.

59. Hoyt, E. P. 1965. *A Short History of Science*. Vol. 1. New York: John Day & Co.

60. Hoyt, E. P. 1975. *Arab Science*. Nash, New York: Thomas Nelson, Inc., pp. 44-46; 127-130.

61. Huff, T. E. 1993. *The Rise of Early Modern Science*. Cambridge Univ. Press.

62. Irving, T. B. 1985. *The Qur'an: First American Version*. Brattleboro, VT: Amana Books.

63. Irving, T. B. 1990. *The End of Islamic Spain*. Cedar Rapids: The Mother Mosque Foundation.

64. Jillispe, Charles C. (ed.). 1970. *Dictionary of Scientific Biography*. New York: Charles Scribner & Sons.

65. Judd, Gerrit. 1966. *A History of Civilization*. London: Collie-Macmillan.

66. Kasir, D. S. 1931. *The Algebra of Omar Khayyam*. New York.

67. Khairullah, A. A. 1940. *Outlines of Arabic Contribution to Medicine and Allied Sciences*. Beirut.

68. Khan Rahman, A. M. 1973. *Muslim Contribution to Science and Culture*. Lahore, Pakistan: Sh. Muhammad Ashraf.

69. King, L. S. 1958. *The Medical World of the Eighteenth Century*. Chicago: Univ. Chicago Press.

70. Lane-Poole, S. 1889. *Story of the Moors in Spain*. New York.

71. Lari Musawi Rukni, S. M. 1980. *Western Civilization Through Muslim Eyes*. Great Britain: F. J. Goulding.

72. Loudon, I. (ed). 1997. *Western Medicine, An Illustrated History*. Oxford: Oxford Univ. Press.

73. Lyons, A. S. and R. J. Petrucelli. 1987. *Medicine: An Illustrated History*. New York: Abradale Press, Harry N. Abrams, Publ. pp. 302-325.

74. Mansfield, Peter. 1976. *The Arab World*. New York: Thomas Y. Crowell.

75. Mathe, Jean. 1980. *The Civilization of Islam*. New York: Crescent Books.

76. Meadows, Jack. 1987. *The Great Scientists*. New York: Oxford Univ. Press.

77. Maududi, S. A. 1980. *Towards Understanding Islam*. Lahore: Idarar Tarjuman Quran.

78. Mirza, M. R. and M. I. Siddiqui (eds). 1986. *Muslim Contribution to Science*. Lahore, Pakistan.

79. Nasr, S. H. 1976. *Islamic Science*. England: World of Islam Festival Publ. Co.

80. Nasr, S. H. 1970. *Science and Civilization in Islam*. New York.

81. Neufeldt, V. and D. B. Guralnik. 1988. *Webster's New World Dictionary*. Cleveland: Webster's New World.

82. Pasha, S. H., Ph.D. 1978-1981. Personal communication and community lectures.

83. Renan, E. 1878. *Miscellany of History and Travel*. Paris.

84. Ronan, C. A. 1982. *Science: Its History and Development Among the World's Culture's*. New York: Hamly Publ. Group, pp. 201-236.

85. Ronchi, Vasco. 1970. *The Nature of Light: An Historical Review*, trans. V. Barocas. Heinemann.

86. Rosenthal, F. 1990. *Science and Medicine in Islam*. Norfolk, VA: Galliard Printers.

87. Sarton, George. 1927. *Introduction to the History of Science*. Vol. 1-2. Baltimore: Williams & Wilkins.

88. Sarton, George. 1952. *A History of Science*. Harvard Univ. Press.

89. Sarton, George. 1962. *Sarton on the History of Science*. Cambridge, Mass: Harvard Univ. Press.

90. Sayili, A. 1981. *The Observatory in Islam and its Place in the General History of the Observatory*. North Stratford, N. H.: Ayer Co. Publ.

91. Scott, S. P. 1904. *History of the Moorish Empire*. London: J. B. Lippincott Co., pp. 490-92.

92. Sedillot, L. A. 1854. *History of the Arabs*. Paris.

93. Shashine, Y. A. 1971. *The Arab Contribution to Medicine*. London: Univ. Essex.

94. Shushtery, A. M. A. 1965. *Outlines of Islamic Culture*. Lahore: Sh. Muhammad Ashraf.

95. Siddiqui, M. R. (no date). *The Contributions of Muslims to Scientific Thought*. Hyderabad, Vol. 14.

96. Singer, Charles. 1959. *A Short History of Scientific Ideas*. Glasgow: Oxford Univ. Press, pp. 137-174.

97. Smith, C. T. 1967. *An Historical Geography of Western Europe Before 1800*. New York: F. A. Praeger.

98. Smith, F. B. 1979. *The People's Health, 1830-1910*. Croom Helm.

99. Stewart, Desmon. 1967. *Early Islam*. Time-Life Books: Morristown, NJ.

100. Syed, Ibrahim. 1989. Islamic Medicine: 1000 years ahead of its time. In: Athar, S. (ed): *Islamic Medicine*. International Publ. Rep. Group: Karachi.

101. Sykes, P. 1934. *A History of Exploration*. pp. 31, 45-51,84,96.

102. Taylor, Walt. 1933. *Arabic Words in English*. Oxford.

103. Turner, H. R. 1995. *Science in Medieval Islam*. Austin: Univ. Texas Press.

104. Tyler, H. W. and R. P. Bigelow. 1939. *A Short History of Science*.

105. Vess, D. M. 1975. *Medical Revolution in France*. Univ. Florida Press.

106. Waalen, J. 1997. Women in medicine: bringing gender issues to the fore. *Journal of American Medical Association*. May 7, pg. 1404.

107. Wallace, M. A. 1959. *The Scientific Methodology of Theodoric of Freiberg*. Fribourg: University Press.

108. Watt, M. 1972. *Influence of Islam on Medieval Europe*. Edinburgh Univ. Press.

109. Welty, P. T. 1985. *Human Expressions: A History of the World*. New York: Harper & Row.

110. The Editors. 1962. *World Book Encyclopedia*. Chicago: Field Enterprises Educ. Corp., Vols. 1, 3, 7, 13, 16.

Index